The Winter Garden

The Winter Garden

Celebrating the forgotten season

Naomi Slade

Contents

Reclaiming the Winter Garden

Making a Winter Garden

Reclaiming *the* Winter Garden

Understanding Winter Gardens

Wintertime gardening is among the dark arts. It is a place that exists on the edge of comprehension: a mysterious and gloomy realm of empty space and quiet enchantments; a slow, gray season of waiting until life and plants return.

Halls may be decked with stalwart evergreens as feasting passes the time and lifts the spirits, and snowdrops are loved and lauded for the hope and light they bring. But until spring floods the world with warmth and flowers, the urge is to withdraw and pay the garden little heed.

Yet a winter garden has fine qualities that should not be dismissed out of hand: the setting sun that floods the monochrome landscape with molten honey and turns seedheads in frost to spun silver; the icy reflections, an inverted world of magic and dreams; the tingling, translucent perfection of small blooms and the crackling warmth of a roaring fire under a crystalline sky, when the connection to the universe is visceral and you can almost smell the stars.

Winter may require a different perspective and a slower rhythm, yet to understand the opportunities and embrace the moment will open a door onto a season that is brilliant, exciting, and rewarding.

A Season of Fortitude

Temperatures drop and the pace may slow, but despite the narrative of horticulture, the garden doesn't ever stop. In fact, quite the reverse is true, as seasonal jobs abound and plants reinvent themselves in all manner of glorious ways.

The idea of "putting the garden to bed for winter" is rather an odd one. Plants don't just decide they are tired one day and stop, and there is no celestial force that arrives to turn out the lights, lock the doors, and declare that the fun is over.

It is a perplexing myth, perhaps recalling the historical dismantling of bedding displays and protecting of tender perennials, while ignoring the gardening that continued regardless. It is illogical and arcane, and, as a concept, it is curiously wasteful.

Total retreat would mean missing the gentle slide of fall into winter: the poetry of a few last leaves clinging to otherwise bare twigs; late-season opportunists such as cyclamen and arums as they spring into life; and the surprising new vistas that are gradually and subtly revealed.

Refreshed by the season, some plants show a new side to their character; suddenly engaging and unexpected; flaunting bright stems or fragrant flowers. Evergreen trees take center stage. Garden structure and hard landscaping reveal their true purpose: a frame for views that are spare and lean, yet still beautiful. And focal points such as containers or sculpture find themselves displayed, as works of art in a gallery.

Even during the shortest, darkest days, life is stirring. Bulbs poke their noses out of the mold, blackbirds gorge on ivy berries, and the baton is passed on in the seasonal relay. Beset by weather and inevitable decay, perennial stems and soft seedheads start to falter, yet quietly and purposefully the garden begins to pick up pace in its rush toward spring.

Countering a season that is prone to moments of cruel cold or trial by deluge, the clear light of an uncertain sun picks out colors that are quietly sophisticated: an elegant palette of sable, ash, and green, against which pale colors glow and shapes and structures seem crisper and more defined.

This, therefore, is not a time to sleep—or, at least, not for long. It is an opportunity to seek inspiration and rejoice in a new and vital form of energy; to make the most of the season and situation, reveling in each small miracle; and to play with color and form, enjoying every possible moment.

Left The winter garden has a quality to it that is unmatched at other times of the year, as it leaves behind the frivolity of summer to reveal both delicacy and fundamental strength.

Where to Begin

Gardens are imagined and made, evolving gently with time to become ever more beautiful. To make the most of the garden in winter, it helps to understand how it works in terms of horticulture and layout, as this will affect the way it is used and appreciated.

Gardens are not abstract—they are personal. And wherever they are and whatever they look like, the fundamental measure of their success lies in their ability to give pleasure.

Approaches vary enormously; some see beauty in a tidy and controlled environment, while others prefer a looser, more naturalistic mood. Both can be a source of pride and nourishment to the soul, flowers for the house, and quiet solitude for morning coffee, but they are fundamentally places for people.

When balmy, floriferous summer is done, it is therefore not enough to just let the leaves fall and live with what is left. This is the point where everything is revealed and every flaw exposed, but to accept this as the status quo and consign the garden to being scruffy and unsatisfactory for four months a year is a waste of potential and a loss of joy.

So take a fresh look at the garden in winter: the plants and the spaces, the nearby buildings and the local architecture. Gaze at the view and absorb the details. Take some photos and sketch the scene.

Think about what you would like to see, recognizing your tastes and preferences, while also taking on board the realities of the season and space. Then add in the factor of time: plants grow, gardens develop, and instant results are a marketing fantasy. For what we see is colored by assumption and familiarity, and when considered closely and objectively, both the problems and the possibilities may not be as they first seem.

The first steps

The realities of wintertime existence are inescapable: it is dark, it is cold, and it may rain a lot. And while fall can feel exciting, with bright gold leaves flying through the air and drifting into crunchy heaps, and a whiff of wood smoke in the air, what follows can seem dreary and dispiriting.

But this can easily be viewed in a different light. While herbaceous perennials collapse, there are plenty of other plants that don't. The trees may be naked, but this simply highlights their potential and an opportunity to create something really good, strong, and interesting from these bare bones.

And although a winter performance would ideally be thought of at the beginning of the gardening journey, we have to start from where we are—not necessarily as garden designers with a blank canvas, but with a space that may already be mature, or in various stages of youth, growth, or decline.

Making the most of what you have got

A good winter garden may be effortlessly strong in other seasons, but the reverse is less likely to be true: a garden that is outstanding in summer is far less likely to put on a stellar performance in winter without some considerable planning behind the scenes. So it is crucial to think about how a planting scheme is going to look when the garden is not at its conventional peak.

Begin by looking at what is there and what is missing or might be added in. Consider whether it works as space in winter or whether it would benefit from rearranging or reconfiguring. Then ask yourself if existing elements might be improved—perhaps there are plants which might be thinned, pruned, or underplanted with something attractive, and unsightly objects that could be removed or screened.

Top right Gardens do not happen by accident; they are spaces that are created and designed.

Bottom right A wildlife-friendly garden does not have to be overgrown; nor does "order" have to be represented by variants of lawn, hedge, and concrete.

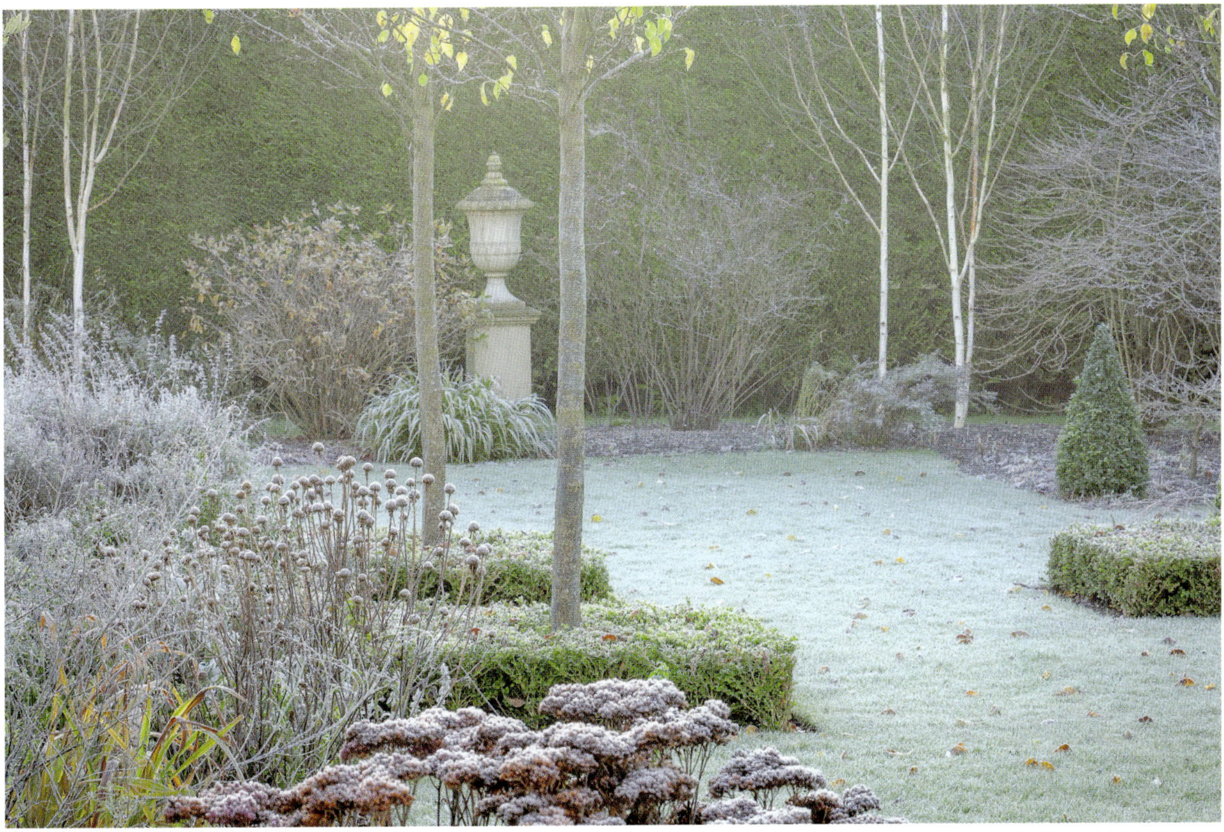

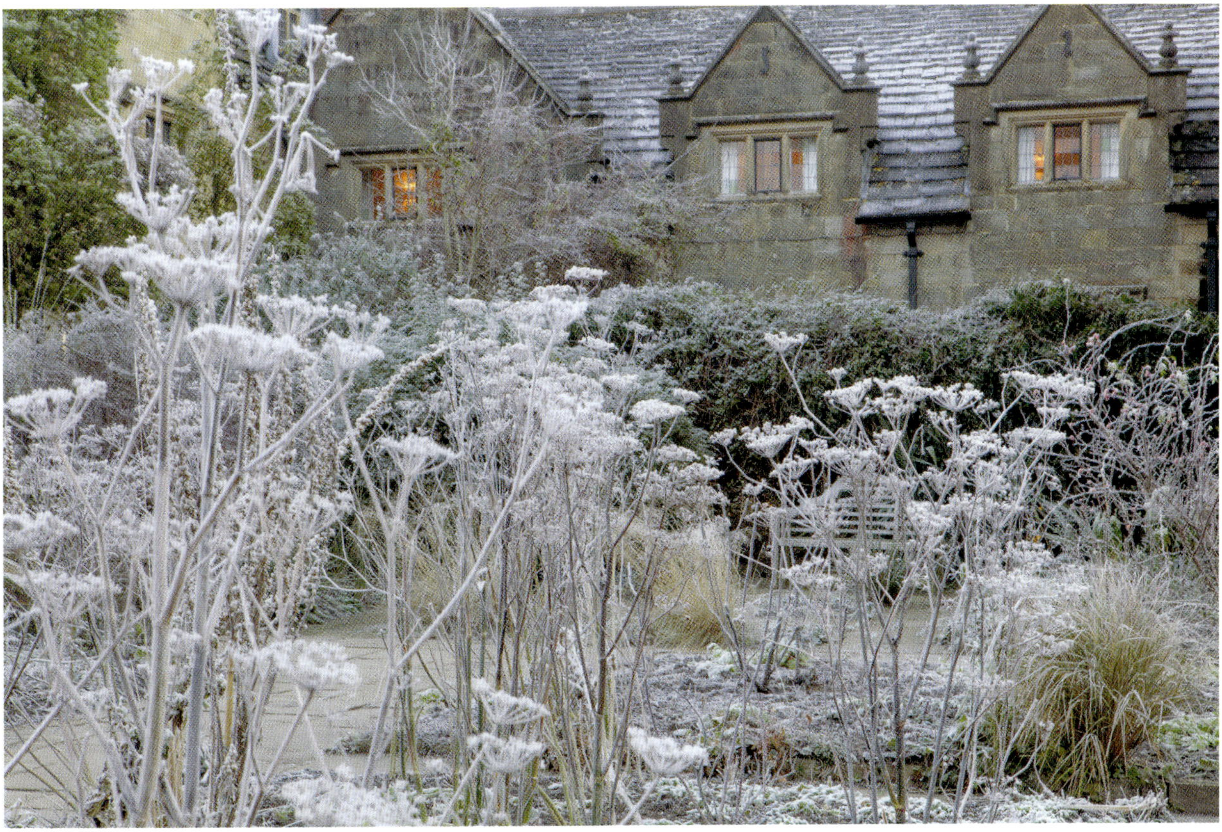

Every garden has positives, so make a list of these and aim to capitalize on them. This could include:

- Favorable microclimates, such as a bank that provides good drainage and suits the plants that like dry feet, or a moist, woodland corner where primroses can thrive.
- Warm walls that provide shelter and retain heat, helping tender plants to overwinter and encouraging flowers into early bloom.
- Large shrubs or trees, that can be employed for their existing height and structure, and that act as a framework around which to build further winter interest.
- A view or focal point that can only be seen and enjoyed at this time of year, connecting the garden and the place with its surroundings to a greater degree, and increasing the size of the space by borrowing visual acreage from the surrounding landscape.

Taking advantage of existing assets will enhance both functionality and experience, while saving money and time on design and build. Starting with at least some mature plants means these can hold the fort while cheaper, smaller additions settle in and catch up. Retaining a connection to the broader surroundings and using the immediate environment wisely, meanwhile, will make for a healthy, lush, and rewarding garden that is a far more enjoyable place in which to spend time.

Garden retail therapy

Having identified areas for improvement or that are in need of a seasonal boost, it makes sense to go shopping—but time of year matters here, too.

For horticultural retail, spring is an important sales period. Fueled by sunlight and cabin fever, the public rushes out to spend money on "brightening things up" and "getting ready for summer."

Hopes and dreams sold by the six-pack are seductive, and it is easy to get overexcited by the color and freshness. But annual bedding and temptingly perfect herbaceous perennials, just getting into their stride,

combine to create a spectacular summer peak that then unceremoniously crashes. Annuals, unless specially grown for seedheads, are ideal for gardeners with commitment issues but not particularly suited to winter, while a reliance on herbaceous perennials has to be carefully considered; otherwise, after the first frost or fall of snow, everything will be flat and brown.

Instead, shopping little and often throughout the year will present plants with different seasonal highlights, which will allow the garden to be interesting for longer. Building a backbone of structural specimens and hunting down plants with several periods of interest, or that are distinctive and unusual, will also help enormously.

The challenges of climate change

The issue of climate change is far-reaching and global, but it increasingly impacts gardeners and gardens, too.

Extreme weather events such as heavy rain and flooding, extended drought, and high winds are becoming more common. In the face of this, planting choices will need to adapt toward more resilient species, while garden features and soil management can be used to mitigate the effects of too much or too little water.

As frosts arrive later, fall is becoming a longer season, blurring into the beginning of chronological winter. There is also an emerging pattern of spring drought. Planting schedules must therefore change to factor this in.

Severe weather is one thing, but consistently mild conditions bring their own problems. As the climate warms, animals and plants are moving northward and becoming active earlier in the year, we may see damage earlier and find different animals feeding on garden plants in future. There may also be an increased risk of fungal problems. This means that accepted techniques and practices may no longer be appropriate, and new strategies will have to be developed.

Right A garden that sits well in its landscape, makes the most of available views, and contains at least some mature plants slides easily and attractively into the winter season.

As we face large-scale global challenges and a profoundly altered world, the collective actions of gardeners count. Together we can harvest rainwater, fix carbon, and make sustainable choices. How we build and landscape can also make a difference. In a natural landscape in Sagaponack, Long Island, New York (see pp.26–27), the architects and landscapers have created a modern building that is sensitive to its local context.

Giving the impression of being nestled in dunes, it forms a "bridge" over a plant-filled bioswale, which both retains the view and allows the house to be resilient in the face of flooding from storms and coastal surges. It is a good example of how we can improve our own environments and mitigate the effects on a local as well as a global basis, as we and our gardens adapt and survive.

In praise of trees

The good qualities of trees are legion, but despite encapsulating every fine thing, from wildlife and heritage value to poetic inspiration, they are received cautiously in domestic gardens.

The reasons for this are varied. Slow-growing and comparatively expensive, a tree is a less attractive impulse purchase than instantly satisfying spring bedding, even though it lasts infinitely longer. There can also be concerns that a tree will get too large, rapidly outgrowing a smaller modern garden and needing to be cut down. This fear is compounded by concerted antitree messaging from insurance companies, which causes people to worry about them far more than is warranted.

But a tree is a glorious thing. According to experts, trees are more important to biodiversity than any other plant form. With a bit of research, there are trees that will contribute permanence, structure, form, and longevity to any space, and they should be embraced as an asset and as a valuable part of the design.

Left Trees come in all shapes and sizes and often respond well to pruning. As a result, they can be used in a garden of any size to provide year-round height and structure.

Looking at Context

To understand the context of a garden is to understand its character on a profound level. It not only influences the planting but it encompasses the geographical attributes, the surrounding scenery, and the potential for problems and challenges. More than this, however, it brings an understanding of the position of the garden in the world; describing the traditions and techniques of the local gardening population, which, in turn, provides an illuminating insight into the strength of its winter performance.

Context is the summation of subtleties. The geographical context of a garden can be as simple as its urban or rural location, but the type of town and the region are important, as are the age, style, and architecture of the buildings nearby. Indeed, location is influential on every scale. A city garden may benefit from the heat island effect, or may be geologically challenged by lack of soil, a post-industrial location, heavy clay, or a high water table. A garden on a hill may have a lovely view, but face howling gales. Parallel plots in Glasgow, Nice, and Copenhagen may all be town gardens in a European city near the sea, but their differences in temperature and precipitation are likely to outweigh their similarities.

With a broader perspective, social history is also important, and ideas vary as to what may be considered established and ordinary. National approaches to international trade are key: the work of the East India Company, for example, running in parallel to the sanctioned piracy of plant hunting, had a huge influence on the composition of Northern European gardens. Added to which are the political and cultural effects of thousands of years of empire, be it Roman, British, Spanish, or Austro-Hungarian.

Each country has its own traditions when it comes to collecting, breeding, and showing plants, and this affects what people buy and grow to this day.

The UK, for example, has a passion for perennials. It still enjoys the Victorian influence regarding ferns and bedding, and the really good collections of shrubs and trees tend to be in older gardens, planted around the turn of the last century. The French gardening tradition, meanwhile, is highly structural, with tightly clipped parterres and trained fruit, and Chinese and Japanese gardens are often stylized and formal.

France excels in breeding and marketing woody shrubs, and these are also popular subjects in the US and Russia. The Dutch and the British love bulbs, but while snowdrops are a winter essential in the UK, they are a minor player in the US. Asia, meanwhile, is not just a source of now-established garden plants, but home to a number of extremely talented modern commercial breeders.

These observations are general, but to some extent the tendency of gardens to underperform in winter is due to their cultural context—an unintentional side effect of not just regional disparities in climate, but countries' differences in fashion and their horticultural marketing spend over the last several hundred years. Without woody shrubs, a winter garden becomes flat. Without bulbs or evergreens, it can seem colorless.

But with the globalization of information, fashion moves faster, enabling ideas to be exchanged, new strengths to be adopted, and holes in the planting palette filled. Social media creates real-time evolution in ideas, driving changes in taste and expectation and challenging the accepted norms, potentially filling any available space, from a balcony pot or an urban courtyard to an expansive country idyll, with interest, inspiration, and color—even in winter.

Right Gardens are influenced by everything from geography to social history. For example, at the National Trust-owned Bodnant Garden in North Wales, UK, part of the winter display is a magnificent collection of North American conifers, which were new and hugely fashionable in the late nineteenth century, when the garden was planted.

Winter Inspiration

Falling in love with wintertime gardening is a momentous event. It may be a rapid, heart-thumping realization that there is a special quality to this time of year, or it may be a slow-burning building of regard and appreciation. In either case, when the charms of the season become impossible to ignore, rejoicing is called for, as it is a small joy that can only make life better.

When the gardening bug bites, the catalyst is usually the brilliance of spring or the fragrant flowers of summer, with winter often coming as an afterthought. New gardens have often marked highs and lows, but with the passage of time, the peaks of interest gradually start to level out, as the borders become full and plants reveal themselves to have a longer performance and broader repertoire than was originally thought.

Winter is by far the best season to see how a garden works and what it is really up to, and taking every opportunity to visit horticultural attractions provides a unique chance to pick up on key details. Larger gardens have often been purposefully planted with areas of winter interest, and run tours to tempt people in, regardless of the season. In the best of these, talented and dedicated head gardeners strive to bring something special and unusual to the experience, showcasing schemes that go far beyond effective but unimaginative combinations of white birch and cornus.

Right It takes time for gardening in winter to become second nature, and for the planting to fill the space. It is very often an exercise in trial and error, in learning from mistakes and having the courage to give it a go. With inspiration from outings and knowledge gained from books, friends, or even social media, the garden will gradually grow and flourish.

Very often, though, the thing that makes the final connection is something that has always been a pleasure: the sun shining through the mist when walking during a sunny, frosty morning, or the snow on pine trees during a winter holiday; the fascination that comes from looking at natural forms, based on spirals and endless repetition, or the childhood story that was so captivating and vivid that each winter brings a secret search for that remembered excitement.

Any of these things can exist independently of an interest in gardening, yet when the world is seen through a gardener's eyes, there is suddenly a new depth and meaning. The experience of winter comes with a new sense of agency, a host of sensory possibilities, and opportunities for creativity and exploration.

Rather than being befuddled by the overwhelming intensity of summer, there is something that is visceral and decisive about winter: you know by gut feeling whether something works or not, and whether it is interesting. Then, secure in that knowledge, it is possible to unpick what is going on and why it appeals.

In this way, you can choose plants that are at their best in the winter season, or take a fresh look at well-known varieties and appreciate their off-season charms, possibly for the first time. It is also an opportunity to walk through a garden, thinking about its sense of space and noticing the surfaces beneath your feet, or the way that an uneven site has been accommodated. It is possible to see the balance between hard surfaces and areas

Top far left Their color may have faded, but airy miscanthus seedheads and sculptural teasels (considered invasive in parts of the US) remain as beautiful as ever.

Top left The mythic qualities of a snowy, frosty woodland instantly transport you to a time of fairy tales by the fire and an imagined land of endless winter.

Bottom left Trimmed beech topiary shapes look almost alien in the dramatic dawn sunlight.

Overleaf This modern house in Sagaponack, Long Island, New York, has been designed to make the most of the rural views from indoors, while maintaining the visual relationships within the existing landscape.

of planting, the first subordinate to the second, but integral to the way that the plants and combinations are experienced.

Preparing a garden for the delights of winter does not have to be expensive, and certainly doesn't have to be high maintenance, but does involve thinking about things in a different way. No garden is too small—or too large—to bring color, texture, and scent into play, while good combinations and repeated use of signature plants or themes can simultaneously take you on a journey and anchor the garden's design.

Building the narrative

A garden is usually thought of as something to be looked at from the house, or somewhere to spend time and immerse yourself in the greenery. There is another aspect to this role, however, and that is to make sense of the local architecture.

A view is reflexive; it works both ways. For all one can look out of a building onto the landscape and appreciate how well the garden connects the two in terms of color, interest, and vistas, surrounding a house with plants and designing a garden to encapsulate it means that the building and its context becomes part of the view for the person looking inward.

At Abbotsford, Scotland (see pp.46–47), the nineteenth-century baronial mansion segues into its rural setting via enclosing walls and waves of increasingly informal planting, but modern buildings also benefit from good and appropriate garden design, even when the architecture is exemplary. The purpose-built home overleaf is bold and striking, yet without the surrounding plants and trees, it would look bald and rather bleak. Here, the intelligent planting allows the building to play visually with levels and perspective, seeming to float as it bridges the grassy ravine that leads off into the distance. Even as late fall fades into winter, the trees embrace the house and the tangled grasses and perennials lend an atavistic quality that connects and settles the building into its environment and gives it a sense of permanence and belonging.

The Garden Experience

Every season has something in which to rejoice, and a particular quality for which it is known. The freshness of spring is utterly distinct from the languor of summer or the spicy crackle of fall. Winter's signature, meanwhile, is invigorating, for this is a season of robustness, known by its ice-bright mornings, buffeting winds, and sharp squalls of cold rain. Yet many people simply shut the door in late fall, abandoning all prospect of going outdoors until the daffodils joyfully trumpet the all clear.

In the face of bracing prospects, we anticipate gloom. We declare the weather dreary and ourselves despondent, then retreat to a comfortable airless place with the intention of holding out until the light returns—which is paradoxical, given the fact that there is plenty to see and do.

The winter garden excels at blowing away the cobwebs and presenting us with something that is not just good but glorious; at creating a confection that will tempt us outside, and that then delivers sufficient enchantment to keep us there. And this is far from being an unrealistic aspiration, for the garden may be lean, and its pleasures fleeting, but the one thing it doesn't need to be is drab.

Making a Space *to* Enjoy

Gardens are about more than just growing things.
They are part of a home and a lifestyle, and no matter
how big or how small, they can be arranged and used in
a way that makes everyday existence more pleasant.

When the weather cools and the pace slows, the keen gardener gladly leaps into action to address any one of a hundred absorbing and necessary horticultural tasks. Yet the outdoor space can still be used for pleasure, relaxation, and entertaining, despite—or even because of—winter.

While days may be shorter they are fresh and reviving, and planning ahead can make all the difference, not just to aesthetics but to health and morale, too. Gardening represents an activity that is good for body and soul. Even pottering around involves bending, stretching, lifting, and light cardiac exercise, so it's not just useful and productive, it also helps to maintain a level of fitness at a somewhat sluggish time of year.

Getting outside into the bright light and fresh air also benefits mental health enormously. Spending time in a natural setting has measurably positive effects on well-being, while activities

such as planting and pruning boost feelings of empowerment and connection with the environment. In an uncertain world, the garden is something that can be influenced and controlled, not only creating a sense of agency but contributing to resilience and self-esteem by way of a small yet positive personal contribution toward combating wider issues such as the climate crisis.

Green fingers notwithstanding, there are a number of things that will almost instantly improve an outdoor area and make it more enjoyable to use. Decluttering works outside as well as inside, and any garden can be made to feel larger and less chaotic by tidying things up.

Getting to grips with seasonal jobs can provide a sense of purpose, but at the same time it helps to have a flexible approach to gardening in winter. On a glorious day, standing in the sunshine and watching the flowers bloom can feel like paradise, but heavy snow or sharp frost requires a different approach in terms of robust planting, factoring in the needs of wildlife, and acknowledging that it can sometimes be best to appreciate the garden from indoors.

Left A garden that is secluded, rich with detail, and reflective of personal tastes can be just as much of a sanctuary as any well-designed room indoors, and because it is constantly changing there is always something fresh and intriguing to look at.

Garden Living

Since the dawn of time, humankind has dealt with months of darkness and cooler weather by celebrating light and striking fires, with festivals and feasting. Whether it is a case of toffee apples for Halloween or fireworks for Diwali, celebrating Thanksgiving, Christmas, New Year, or Imbolc, there is always an excuse to live it up, and the party is nearly always more fun when it is held outside.

Short days and chilly breezes do not have to mean that the outdoor space is off-limits, and a fairly modest investment in shelter and warmth can keep the garden not just usable, but an actively enjoyable place to be.

When the house is packed and the merrymaking is in full swing, the doors can be thrown open for the guests to spill outside and enjoy the fresh air. A few comfy chairs in a greenhouse, conservatory, or outbuilding will instantly add to the living space, and provide an alternative location to hang out. And, if the weather is not too inclement, entertaining can still take place under the stars.

A firepit, some seating, and a few blankets to hand make it easy to snuggle up with a glass of fizz, while twinkling lights add extra atmosphere after dark. Grilling and campfire cookouts are fun all year round, outdoor kitchens continue to earn their place, and dining areas decked with candles and greenery make for an enjoyable twist to intimate dinners.

Existing garden buildings can be adapted or bedecked for the season, and will provide a bolthole should the heavens open. Special birthdays, anniversaries, and celebrations, meanwhile, need not be confined to the summer garden. Temporary solutions exist to protect lawns from footfall, while dinner and dancing can take place in a pavilion, hired for the night and lit by lanterns.

When the party is over, permanent structures can be used as a quiet personal retreat—somewhere to have a cup of coffee and a catch-up, as a pleasant change of scenery, or a secluded spot in which to write letters, read books, or have a private phone call—particularly if any stray house guests should remain.

Appreciating the winter garden

Alongside tackling the practical jobs around the garden and using it for occasional gatherings, it is nice simply to be outdoors, so take time to enjoy the moment.

Keep warm and dry and shrug off bad weather by muffling up in waterproof clothes and retreating to a covered place. Garden rooms are great if space and budget allows, but an arbor seat does a fine job of keeping off the drizzle while you enjoy a mood-boosting hit of natural light.

Organize the view by arranging the planting in a scenic fashion to delight the eye and stimulate the senses. A cluster of colorful pots provides instant interest, as do bulbs and hellebores, sweetly scented flowering shrubs, or a witch hazel (*Hamamelis*) sited where the orange-peel flowers will catch the light. When the weather is too dreary to go out at all, there are a number of ways to enjoy the garden from inside (see p.39).

Go for a walk. The shortest stroll can be refreshing, and even a simple loop of the garden can become a journey. Keep the planting coherent by repeating key elements, such as a particular shrub or evergreen, and linger long enough to notice the naturally evolving scheme of interest that changes the mood daily.

Top right In winter, a shed can be repurposed as a place to sit out and take a quiet break away from the hubbub of everyday life.

Bottom right Invite people in with a welcoming entrance to the house and garden.

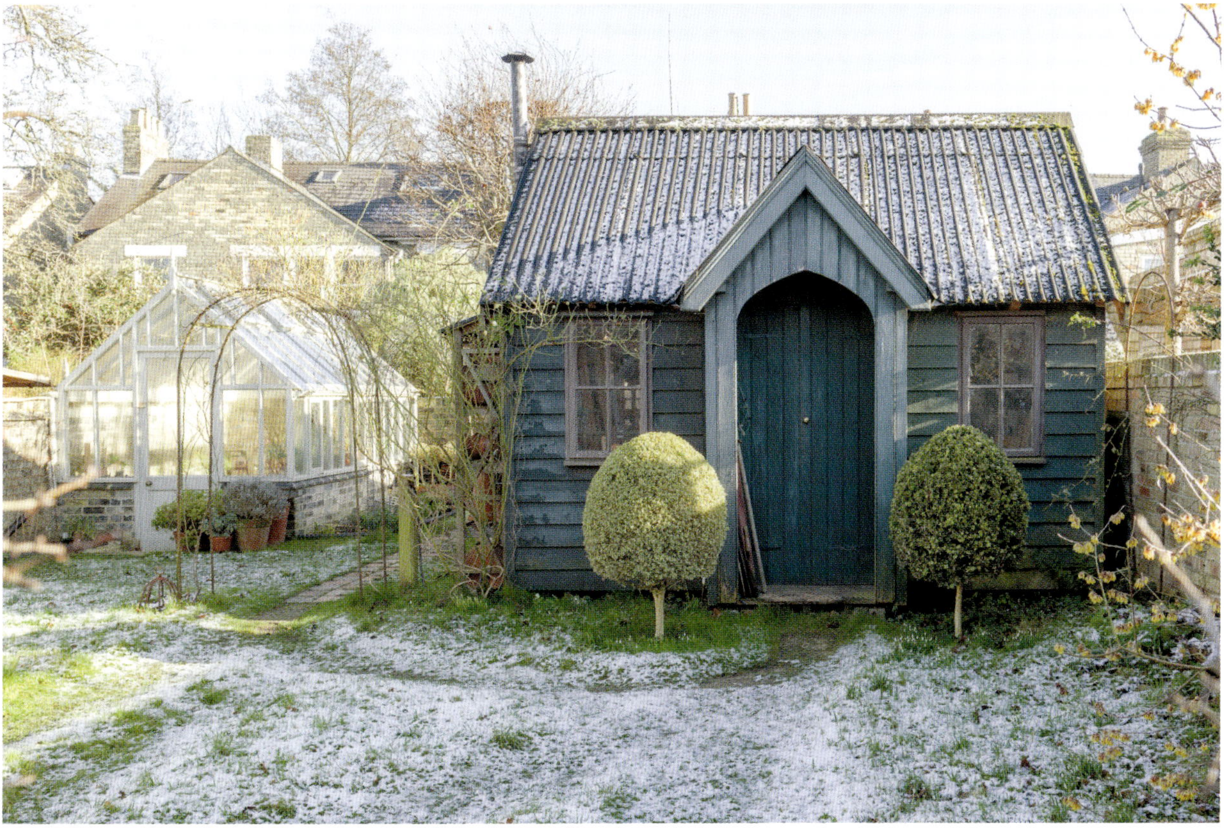

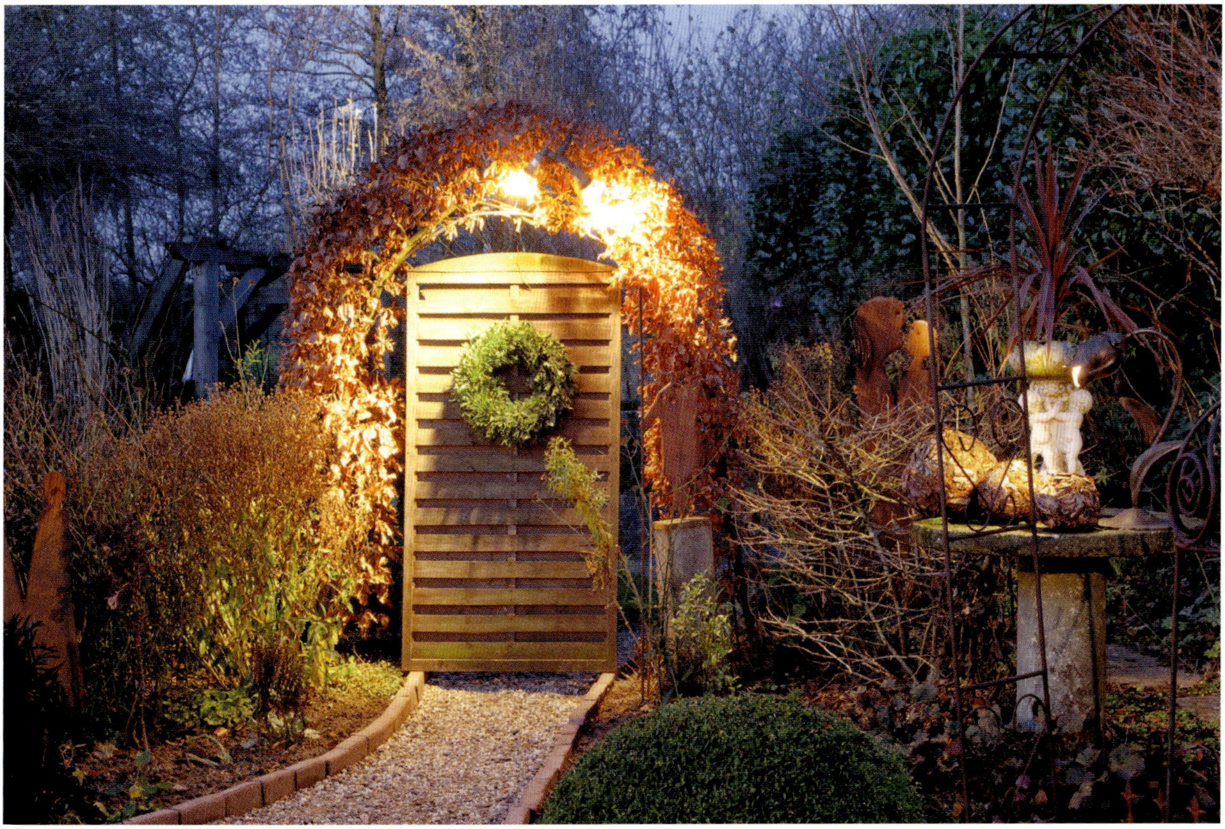

Practical Jobs *for the* Winter Season

Working outdoors in winter is good for the mind, body, and soul. It offers fresh air, daylight, and a bit of exercise while getting ahead with tidying, pruning, and dividing, so that by spring most of the hard work is already done. It is also a time to think hard about putting major changes into place, while the garden is at its most lean.

Things to do and plants to grow

Bulbs: planting up containers and spreads of bulbs in fall and early winter ensures there will soon be flowers to enjoy. It can also help to link the winter and spring gardens, so there is not an awkward pause. Feed bulbs in containers to keep them going until the following year (see p.98).

Tender perennials: some plants are borderline hardy and these can be lifted and brought into a greenhouse, or moved under cover wholesale if grown in a pot. Alternatively, if the soil is sufficiently well-drained, they can be left in situ, with the crowns protected with a thick layer of mulch or insurance cuttings taken in late summer and overwintered inside.

Fruit trees and other woody plants: winter pruning encourages new growth that will produce fruit and flowers, and helps to maintain an attractive shape (see p.130). However, if a plant is doing well or is flowering consistently without intervention, annual pruning may not be needed, while very mature specimens may not need pruning at all.

Soil care: using organic matter as a mulch does not just make the garden look tidy and cared for, but it also improves everything from water retention and drainage, to increasing the availability of nutrients and supporting microfauna. Add spent potting mix, leaf mold, or chipped bark to the surface before the cold sets in and let the weather and worms do their work.

Sowing seeds: in early and late winter, seeds can be sown for a crop of annual flowers such as sweet peas or *Ammi majus*. Chiles and tomatoes can be sown in early spring, to fruit in the summer (see p.165).

Bare roots: in fall and winter, a range of perennials, shrubs, and trees are available to purchase as dormant, bare-root specimens. Often cheaper to buy and ship than potted plants, these may need less maintenance too, as they are inactive when planted and the ground is already moist.

Tidy up: it is much easier to relax and enjoy your garden when it doesn't present you with niggling jobs to do, so take some time to sort things out to your liking. Stack empty pots in an inconspicuous corner, compost dead plants and debris, and make sure that everything looks as pleasing as possible.

Notes and inspiration: rather than trying to keep track of thoughts and ideas from season to season, invest in a notebook. Use this to list key plants and new ideas, identify areas of interest, and remember where the gaps are, as well as sketching plans and noting the location of dormant perennials. Creating a record as you go along is far easier than trying to remember everything later.

Clockwise from top left Bare-root *Ligustrum ovalifolium* in a bucket, waiting for planting. When temperatures start to dip, move tender plants such as echeverias and aeoniums into a light and sheltered place; a cool greenhouse is ideal. Pots of bulbs such as *Eranthis* and snowdrops are a cheerful sight on a snowy day.

Coping with cold

While moderate cold can be managed fairly easily, when the temperature drops sharply or there is a prolonged freeze, the garden can start to struggle. Some measures for dealing with cold are straightforward: avoid walking on frozen lawns, and make sure that there is water available for wildlife to drink. Tender plants, such as banana palms or tree ferns, can be wrapped in sacking and insulated with straw, while a blanket thrown over lower-growing specimens will hold off the worst of the frost.

Once temperatures reach 23°F (−5°C) or lower, however, cold becomes a serious problem not just for more delicate subjects, but for those that are usually considered hardy as well. Containerized plants are particularly vulnerable as the pots can freeze solid, so if very cold weather is forecast, move anything remotely tender into a frost-free place such as a conservatory or unheated spare room.

In this state of gardening emergency, aesthetics are not relevant and light is less important than survival, so cluster small pots of perennials and bulbs together and stack them next to a wall. Insulate them with cardboard, or surround them with bags of potting mix to make a bunker, which can then be topped with more cardboard, or an old blanket or duvet.

Snow is a different issue altogether. When heavy or sticky, it may flatten perennials and weigh down branches, potentially breaking them; knocking it off with a broom will relieve the weight. Evergreens are particularly susceptible, but binding up the branches, where possible, will provide a temporary solution.

Yet a blanket of snow will also shield the ground and protect the plants beneath from severe air frosts. As it melts, meanwhile, it gently irrigates the underlying soil, so it can actually be an asset. In the wild, many plants, particularly alpines, are adapted to benefit from the insulating properties of snow and its role as a seasonal water source.

Left Frost effects are pretty, but when the weather gets really cold, steps need to be taken to protect the garden, and there may be some losses.

A Room with a View

Sturdy of spirit and optimistic of outlook, gardeners are a naturally pragmatic breed, yet for every day of golden sunshine, azure skies, and exquisite frost effects, there are likely to be ten more where chilly, torrential rain is driven by a stiff breeze. It is on days like these that even the hardy souls who are much invigorated by bracing weather are more inclined to stay indoors.

Winter has predictable downsides, so it is worth planning ahead to make sure that the parts of the garden that can be seen from the windows are as delightful as possible a lovely view can be a huge boost and revive any sagging spirits.

Sprinkle solar lights around dark and twiggy corners, and adorn potted trees with tasty bird-treats to encourage visitors. Deck doors with natural wreaths and welcome the world in by creating a glorious entrance.

Bring fresh plants closer to the house by packing window boxes with color to brighten up gray days, and planting the nearest borders with winter in mind. A display of containers in the foreground will help to distract attention from failings further away, so gather together pots from around the garden and arrange them outside the window to form a cluster of all the things that are still looking good.

This display can then be brightened with seasonal bedding and supported by containers of bulbs that have been planted up in fall and left in an unobtrusive corner until the buds appear. As the interest fades, new containers can be added in, then replaced in their turn. This creates a cycle of interest and allows the display to remain fresh, which is particularly effective in smaller spaces where there may be less to work with in the surrounding landscape.

Connecting the interior and exterior with a decorative color theme or choosing plants that pick up the hue of a key feature can be effective, and the garden can also be united with the house by bringing a little bit of it indoors and making it part of that environment, too. Cutting an armful of twigs, greenery, and whatever flowers present themselves and arranging them in a vase near a window will create a link with the immediate landscape and satisfy personal tastes in a way that a bunch of shop-bought blooms can never do (see pp.168–185).

Fragrance is one of the more surprising treasures of the winter garden, and picking perfumed plants is an excellent way to enjoy them in cold weather. Scent is made up of a range of volatile chemical compounds that are released by the plant and evaporate into the air. This process is inhibited by cold, as the aromatic compounds don't evaporate as fast. A frosty morning smells fresh and clean for the same reason. Warming the plants up activates a fuller perfume profile, and indoors, with no wind to blow them away, the pleasant floral or piney scents will hang in the air for longer.

The weather may be atrocious, or life may just be too busy to head outside, but the garden is still part of the home. A little thought and planning means that it can be treated as a living picture, enjoyed for its resilience and in all weathers as it gently evolves, while a fresh and sweetly fragrant posy on the table will improve that connection further, and make the very most of all the pleasures that are still available.

Far left Large windows are particularly effective at connecting a building with its surroundings.

Left Planting for an attractive view means that the garden can be enjoyed even when the weather is dire.

Welcoming Wildlife

While aesthetics and usability play a significant part in our own enjoyment of the winter garden, it is worth bearing in mind that there are a lot of other creatures that live there too. What we do can be actively beneficial or unnecessarily destructive, but, ultimately, choosing to support nature and prioritize the environment will also make our own world a better place.

With short, cool days, the natural world slows, and for many animals the focus shifts toward sleep or survival. Rather than being able to spend the long hours of daylight foraging, finding somewhere peaceful, dry, and undisturbed to rest becomes essential. Animals and birds will also be hungrier during periods of cold weather, so nutritious food and nearby sources of water are crucial too. Planning for wildlife is, therefore, as important as ever.

Fortunately, an environmentally sensitive approach and an aesthetically pleasing garden are not mutually exclusive, and many sympathetic gardening practices work particularly well in winter.

Layering up shrubs provides us with visual depth and the birds with roosting spaces, while a dense canopy with overhanging branches will keep temperatures higher than the surrounding landscape, helping to fend off the frost, which will benefit plants and animals alike. Walls thick with ivy and tangled with climbers will shelter overwintering butterflies, alongside the aphid-loving ladybirds and lacewings that are so useful in spring. Leaving stems standing protects the crowns of dormant perennials and provides holes for insects to hibernate, while raking leaves off the lawn and piling them in a corner with twigs and brash will make a cosy retreat for hedgehogs, slowworms, and beetles.

Because birds don't hibernate, they need a constant supply of food and water, so leave seedheads in situ and plant berry-rich shrubs. Putting up feeders and a bird table just outside the window and creating a bird tree, festooned with edible baubles, will provide additional nourishment.

When it comes to insulation, clean, fluffy feathers are best, so birds need water to bathe in as well as to drink. Keeping the birdbath full is important, and making sure that ponds and pools are not fully frozen over—and either melting a hole in the ice or putting out extra water if they are—will be a lifeline to a range of thirsty animals and insects.

Some winter days can be positively balmy, however, and in periods of good weather, hibernating insects and animals may emerge to top up their reserves. Ivy flowers provide late nectar for bees and hoverflies, and early-flowering bulbs and shrubs such as *Lonicera fragrantissima* and *Mahonia* also support the bees with nectar on days when it is warm enough for them to fly.

A number of wildlife-friendly features are actively good-looking, too. A decent-sized bug hotel, made up of lots of different materials such as pine cones, pieces of bark, twigs, leaves, and hollow stems, all packed into a wooden container, can be an attractive focal point, while heaps of mossy logs arranged organically, or stacked in a pyramid until they decay, are naturally sculptural. In larger gardens, meanwhile, standing deadwood and old stumps provide excellent habitat.

Top right Seedheads that are allowed to stand into the winter months offer rich pickings to the local birdlife.

Top far right Leaving a stack of logs in the corner creates nooks and crannies for animals and insects to overwinter in, and also provides a stable, long-term food supply for beetle larvae and other detritivores.

Bottom right On warmer days bees will wake up and go foraging. In addition to shrubs such as *Lonicera fragrantissima*, planting crocuses and snowdrops will ensure that there is plenty of nectar and pollen to go around.

Green Living

Gardens can make a real and positive contribution to the environment, acting as carbon sinks, wildlife reserves, and water-management systems. Particularly in cities and other densely populated areas, the way we plant and care for our gardens can have a substantial impact on the world around us.

The simple action of growing plants can improve the local environment in a number of important ways. Not only do plants capture carbon and release oxygen as part of the process of photosynthesis, but their dense foliage slows air flow and helps filter out particulate pollution—even in winter, if they are evergreen. Planted areas are vital in managing water, meanwhile, and the type of planting and the way it is cared for also has an impact.

Carbon

Soils are an extraordinarily effective carbon sink and, as gardeners, we can capitalize on this. Digging and disturbing the soil releases carbon back into the air, while mulching and planting for the long term enriches the soil with organic matter. This not only improves the soil structure and makes it more effective at soaking up rainwater, but some of that carbon is also locked underground where it is stabilized through its associations with a range of microbes and minerals.

Right Trees and shrubs drop twigs and leaf litter onto the soil below; once incorporated, it will improve soil structure and water penetration, and some of the carbon will be sequestered underground. Dense planting reduces the speed at which water runs off the surface, meanwhile, and evergreen foliage helps to clean the air by trapping pollution.

Leaf mold, well-rotted chipped bark, and spent potting mix are all good mulches, and they can be added at any point as they break down slowly and are relatively low in nutrients. Manure is a good source of organic matter, but it also tends to be rich in nitrogen. This is fine if the aim is to produce lots of lush foliage, but the plants won't take the nitrogen up if they are not actively growing. The risk, therefore, is that nutrients that are applied to dormant plants sit around until they get washed away by a downpour, when they may cause problems elsewhere. To avoid this, it is better to spread manure at the end of the winter season, as the plants start to grow.

The other way to trap carbon in the garden is to maximize the permanent planting, particularly of large, woody, carbon-dense specimens such as shrubs and trees. These should be as long-lived as possible as they will continue to grow and fix carbon throughout their lifespan, and even when they are dead much of the embodied carbon remains beneath the ground, within the root structures and humus.

Water

Winter weather tends to be generally wet, but with extreme weather events on the rise due to a changing climate, heavy rain and flooding has become more likely. The garden can help with water management in a number of ways. Trapping water and slowing its flow helps to even out the peaks and troughs in availability and improves infiltration into the soil.

In periods of heavy rain, good-quality soils with plenty of air pockets act as a more efficient sponge than compacted soils, because the air spaces fill up with water. Areas of planting will also trap the water and slow the rate at which it runs off. The denser the plant growth, the better a barrier it creates and the more time there is for the water to sink into the ground, which is a strong argument for leaving lawns long and border perennials standing.

In combination, this not only reduces the risk of local flooding, but protects soils from being washed away and nutrients from being flushed out and ending up in watercourses, potentially causing uncontrolled algal growth. Instead, the water replenishes the reserves within the soil, meaning that it is available for the plants when they start growing again.

Management and conservation

Alongside extreme rainfall, our increasingly erratic weather and the climate crisis means that periods of drought are also more likely, even in winter. Harvesting and conserving water is, therefore, going to be essential.

Installing rain barrels that are fed from roofs and gutters avoids wasting expensive, carbon-intensive drinking water on the garden, as well as reducing the volume of rainfall that runs into the sewers. Once collected, the rainwater can be used for watering pots, the contents of the greenhouse, and houseplants, all of which are likely to need a drink over the winter period. This is preferable for those species that resent alkaline tap water and it also acts as a general insurance against periods of dryness in spring.

A number of design features can be deliberately used to trap surface water. As discussed above, areas of dense planting and permeable paving solutions are useful, but even in quite a small garden, a dry riverbed and occasional pool can be installed, to act as a sump. This will fill up during a downpour, trapping at least some of the sudden influx of water, which can then gradually soak away over time.

Top right Managing water so that it moves slowly through the landscape reduces the chance of local flooding, and it is also enjoyable to look at.

Top far right Grasslands are underappreciated for their ability to lock up carbon, and when turf is left a little shaggy they can absorb a lot of rainwater, too.

Bottom right A beautiful garden with substantial borders, varied permanent planting, a lightly managed lawn, and a pond will encourage wildlife and help the environment.

Getting Out and About

One of the greatest pleasures of gardening lies in leaving one's own efforts behind and seeking inspiration elsewhere. While some gardens may shut for winter, others are increasingly staying open to capitalize on visitors' growing enthusiasm for year-round interest and color.

At this quiet time of year when the garden at home is relatively undemanding, taking time out to see other gardens can reap great rewards. Discovering new plants and fresh ideas is the best way to take our own gardens to the next level, especially if there are areas that are revealing themselves to be problematic.

When faced with a plot which is gappy, thin, or just plain boring, taking a voyage of discovery and checking out seasonal hotspots—and the nurseries that go with them—may be just the solution. With new plants to consider and ideas to adopt, keeping a notebook and camera phone to hand is essential. If there is a particular area of the garden at home that needs improvement, meanwhile, it is a good idea to take a photo of it before you leave, and if you are in pursuit of specific plants or accessories, having a budget in mind is wise.

Inspiration lies in real gardens, stripped back and displaying everything that is best about the season. Even grand creations, which may be utterly beyond scope in terms of size and cost, will always offer something that can be adapted, used, and enjoyed.

Left Abbotsford Castle in the Scottish borders was the home of nineteenth-century novelist Sir Walter Scott, who designed the formal walled gardens to make a transition between the luxurious interior and the wider landscape. Laid out in the Regency style, its three garden rooms are full of detail and interest, and would originally have been filled with the latest plants from all around the world.

Winter Garden Design

Whether designing a garden from scratch or improving an existing arrangement, winter represents an opportunity to build something wonderful, from the ground up.

It is a time to embrace the garden's minimalism, revel in texture and structure, and admire the resilience of the season. Take delight in the quality of light that is exquisitely subtle, and the plants that boldly thrive where others fade and crumple.

From a design perspective, winter is an uncompromising time of year: there is nowhere to hide and there are no distracting flowers or screens of leaves to conceal flaws and gaps. Either the garden is good, robust, and engaging, or it is not.

The landscape is pared back to bones and dark earth; the garden is stripped down to its fundamentals, battered by rain, torn by wind, and gnawed at by frost. Yet there are elements of spine-tingling glory: bare perfection in the first rays of a luminous dawn; an exultation of crocuses; splashes of color that are dazzling and insouciant; and fragrant flowers in the darkness. Sweet rewards, made sweeter still by the fact that they are unexpected and hard won.

The Changing View

As seasons shift, the effects on a planting scheme
can vary wildly. Summer and winter are often seen as
polar opposites, yet to write off the colder months as dull
is unnecessarily simplistic, and it is something
that can also be easily avoided.

As time goes by and the year waxes and wanes, the view from the window evolves. While the bedrock of nearby buildings, distant hills, and mature trees remains more or less static, there are substantial changes in what is hidden and revealed, and what will catch the eye.

A bonny and bodacious shrub border can find itself thin and ignominiously straggly as winter approaches or it can reinvent itself with a burst of berry and bloom. A shady copse can be transformed into a winter fairy tale, the slender branches reaching toward a blue sky, swathes of bulbs covering the ground and a touch of snow melting in the sunshine, while a distant church spire becomes part of a borrowed view.

In garden design, therefore, it makes sense to think of winter first. Meditate on that season in particular, and build on what works.

When getting dressed, underwear comes first, and to start designing a garden when growth is at its peak is rather like donning a nice coat for an outing, then belatedly realizing that the foundation garments that ensure both modesty and a good fit have been forgotten.

In winter all is visible, as it will be again, so at the very least, imagine how the space will look and draw a sketch. Consider framing and layout; views, and screens. Think about winter color, structure, and interest; devise solutions and preempt gaps, so that when the leaves fall you are ahead of the game. Then knit it into the other seasons for a sense of continuity. Some retrofitting of structure and planting for interest will almost inevitably be required, but with a garden that is mature to start with, the results can be pleasingly rapid.

Gardens are always a process of iteration and discovery and, at best, they are composed and edited like a work of art. In a masterful oil painting, the pigment is gradually built up, layer on layer, and the final impression is due to the accumulated effect of thousands of brushstrokes and blended shades. A garden is likewise finely tuned but it can, and must, constantly be refined and added to, in the endless quest for perfection.

Left Planting that factors in the changing seasons and works with the surroundings can result in a garden that is still thoughtful, elegant, and dreamy, well into winter.

The Fundamentals

A garden has a range of jobs to do, and it is worth considering at an early point whether it performs effectively as somewhere that is pleasant, safe, and practical in which to spend time.

A good garden will tell a story, but alongside this narrative, it should sit comfortably within its surroundings and reflect the needs and desires of the people who will use it. It helps, therefore, to interrogate all aspects of the space, and rate its functionality overall.

Start by analyzing the existing garden arrangement:

- How big is the space, and how does it relate to its surroundings in terms of scenery or nearby buildings? Does it have a sense of place?
- Does it connect to the wider landscape? If there are views, are they an asset, or does the garden look out onto something ugly that would benefit from screening?
- Does it perform the role it needs to? Is it safe to navigate or are there steep slopes, uneven steps, or sharp drops? Is there any existing seating, storage, or shelter?
- How do you move around the garden? Is it practical and does it flow?
- What planting is there already? Are there mature trees or specimens that could be capitalized on?

Having assessed what is present, the next thing is to consider how this can be added to and improved upon.

Boundaries and screens: the edges of a space can have a significant impact on the feel of the garden, particularly in a season when the connection to its surroundings is stronger, and the neighbors are more visible, than at other times of year. Designers often talk of blurring boundaries and borrowing the view to make the garden look bigger, but in winter this may not be completely straightforward. New views may appear, but so may eyesores, and once the leaves have fallen the garden can find itself more overlooked.

Layout: a garden should be arranged in a practical manner, so that it is easy to move around and access features such as sheds and compost bins that are in regular use. But layout can also have an enormous effect on the way a garden is experienced. Its sense of journey, potential for surprise, and its aesthetic effect can be improved by minimizing clutter and arranging the component elements so that they have space to breathe.

Materials: hard landscaping is a prominent part of the winter scenery, and the overall impression is improved if the various elements are unified and pulling their weight. Using a limited number of materials and echoing the local vernacular of stone, brick, or wood improves cohesion and ensures that the garden sits well in its environment. Smaller gardens, meanwhile, might echo the theme of an interior, especially if the house has large windows or French doors.

Habitability: this is a rather intangible quality, as what constitutes habitable varies according to taste and circumstance. Fundamentally, though, the garden needs to be waiting to be used and enjoyed, presenting as many opportunities and as few barriers as possible. It also needs to enable ordinary tasks to be undertaken with a minimum of fuss. Taking out the garbage cans may be a chore, but it doesn't need to be an exercise in navigation and hazard avoidance as well.

Top right Large windows maximize the view from inside to out, connecting the garden to the house and enabling it to be enjoyed even when the weather is too cold to sit outside.

Bottom right Considering the balance between the planting and the landscaping means that no single element will come to dominate.

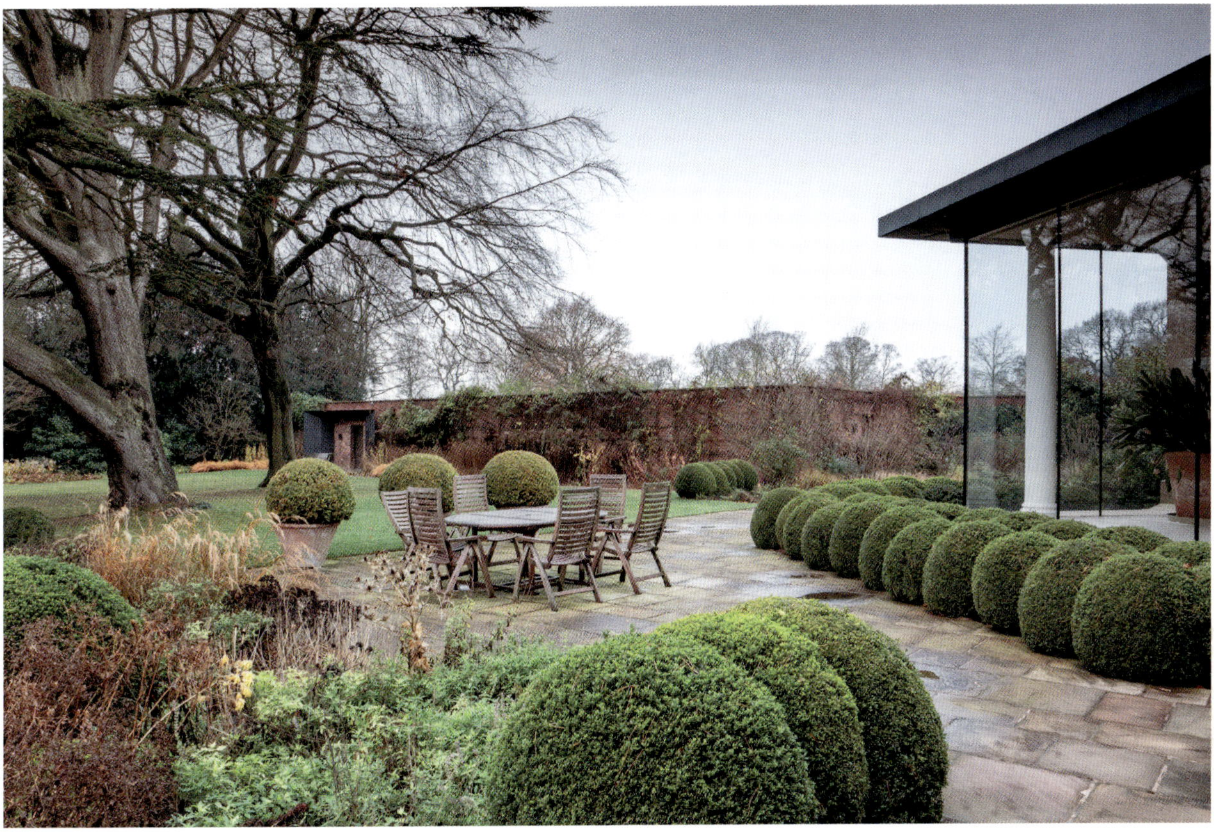

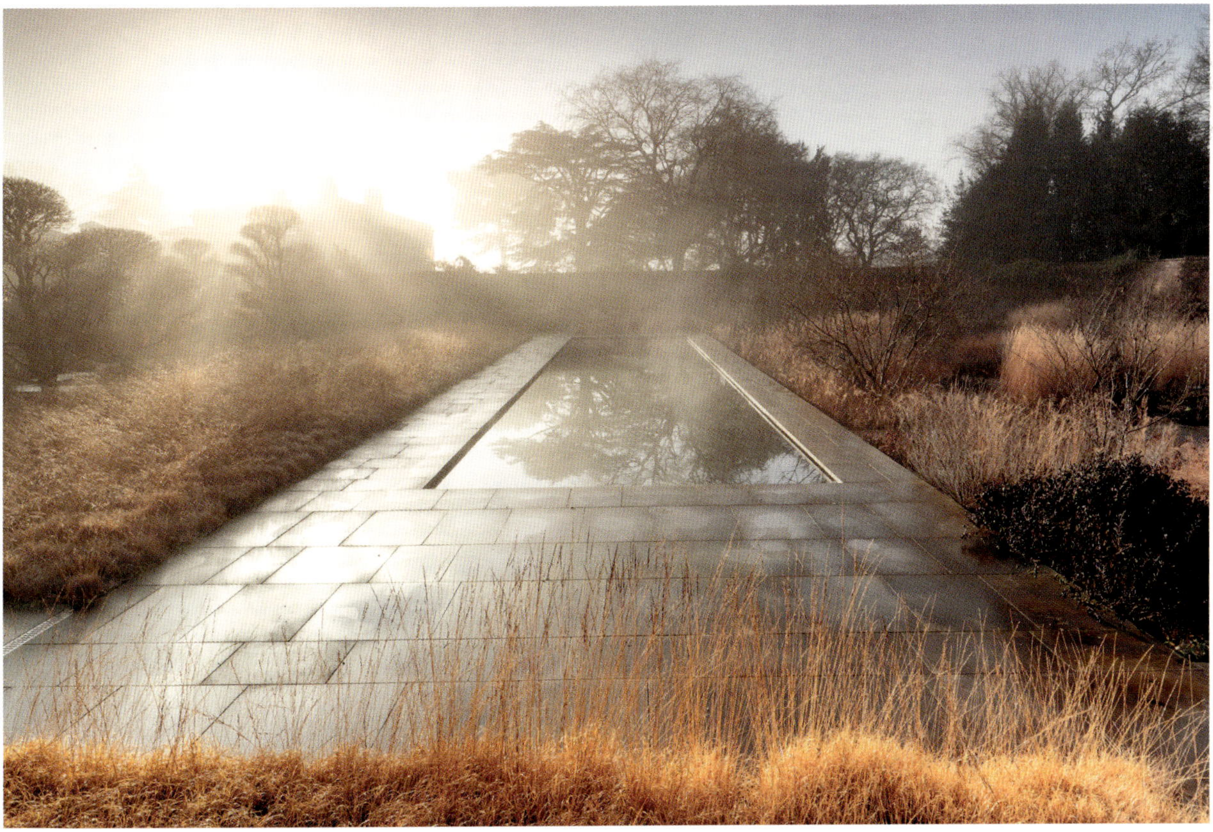

Key Elements *to* Designing a Winter Garden

A balanced composition and a congenial outlook help the garden earn its keep in winter. Form must be good and there needs to be height and color. Simplicity and repetition are pleasing, yet there must be enough interest and variety to hold the attention. It is an art, but one that is underpinned by a few simple specifics.

When designing a garden to embrace and manage the minimalism of winter, there are a number of points to address. The space must be considered in terms of its shape, size, and aspect, along with environmental factors such as climate and prevailing wind. Take into account desirable plants or attractive features, together with issues such as prominent garbage-can storage or neighbors' windows that overlook all or part of the area.

It helps to sketch a diagram of the garden, indicating existing elements, together with the main view from the house. This provides a framework for improvement and changes, and makes it easier to think about design ideas in context.

Top far left A meandering path sets out on a journey through a forest of colored stems.

Top left Judicious screening can be used to frame a focal point while also using the principles of "hide and reveal" to create a sense of mystery.

Bottom left In this layered and symmetrical view, the path is flanked by yew balls and hedging, which draws the eye toward the tempting gap in the wall of clipped beech, and the taller trees beyond.

Height: this is important for filling and managing the space. Think about vertical accents and where they would be desirable. Consider the role of trees in screening and how they might also provide shelter, shade, or a sense of being enclosed.

Repetition: when many different elements are mixed together they often look haphazard and bitty, but repeating a particular plant, color, or shape will add strength to the design and keep it grounded.

Pace and journey: linked to repetition, this is about "stop," "go," and "what's next?" It can be achieved by multiples, such as three birch trees or five topiary cones, or a repeated theme such as a shape or a pop of color to lead the eye along. Upright or spherical forms can be used as "full stops" to indicate the end of one theme and the beginning of something new, while a change of pace or sudden non-sequitur can have a look-at-me quality. Even the width of a path and the materials used can affect the speed of travel and overall feel.

Depth and distance: using screens and voids to break up the space means that not all the garden is visible at once, concealing and revealing different areas in turn to create a sense of anticipation and surprise. Layers of planting add depth, particularly if the boundaries are obscured, and the different elements can act in concert, with new points of interest emerging as the season waxes and wanes.

Framing: when a view or focal point is framed, it adds theater and intensity to a design. The uprights of a door or pergola, or a gap between the trees, will lead the eye to the object of interest, providing something to get excited about, and this elevates the experience and gives the design real purpose.

Time: seasonal fluctuations aside, gardens operate in three dimensions of space and a fourth dimension of time. Plants grow and alter the landscape around them, so visualize how big they will be ten years hence—or even longer—and give them enough space.

Working with proportion

In any design, good proportions are essential and when an arrangement of plants or a structure seems particularly agreeable, it is often based on the Golden Ratio, 1:1.648—also known as the divine proportion. As with the simplified "rule of thirds," its principles are frequently used by photographers to compose pictures, where the subject takes up a third of the space and the context takes up the remaining two thirds.

In geometry, this is represented by an arrangement of rectangles and squares, set into each other so they get smaller and smaller. This can be overlaid with a wide-mouthed spiral to demonstrate an infinite regression.

For reasons that are not fully known, these proportions unfailingly bring harmony and balance. Used by artists such as Leonardo da Vinci and seen in the architecture of the Parthenon in Athens, the golden rectangle works just as well when planning flowerbeds and terraces, or dividing up a garden using paths. With structural elements particularly prominent in winter, this is when we need strong, coherent, and pleasing geometric principles most.

The Golden Ratio is related to the Fibonacci sequence, which governs spirals and patterns throughout the natural world. The sequence is achieved by adding each number to the one before, so you have: 0, 1, 1, 2, 3, 5, 8, 13 and so on. The ratio between two successive numbers becomes closer to the golden ratio the bigger the numbers get, and this helps to explain why plants in groups of three or five always look good, and so do arrangements grouped by height, with, say, one short tree and two taller ones.

Right Certain kinds of repetition or the proportions in a shape are inherently pleasing to the human eye. These are very often related to the numbers in the Fibonacci sequence and the rule of thirds, which is a simplified form of the Golden Ratio.

Effective Planting

In winter as at other times, plants are at the heart of a garden. They provide colour, movement, and interest, and can be combined to exquisite effect, holding the frame of the space and creating a picture that gently evolves.

Endlessly appealing and enormously adaptable, plants improve our lives no end. Not just easy on the eye, they moderate our environment and attract wildlife, while also enabling us to soften challenging views and cheaply hide unattractive garden features that might not be prioritized for immediate removal. It makes sense, therefore, to choose the best plants for the situation and task in hand, and to combine them as effectively as possible.

Spacing is important in giving plants the room they need to perform. Hedging plants should be planted fairly close together, a shrub border should be cohesive but not overcrowded, and a specimen needs enough space around it to make it stand out.

Contrast colors for impact. Orange stems pop next to purple flowers; white tree trunks can be offset with dark yew. Black *Ophiopogon planiscapus* 'Nigrescens' makes yellow foliage zing, while intensely blue *Scilla siberica* brings acid green *Cornus sericea* 'Flaviramea' alive.

Introduce layers to add richness, depth, and density. Start with trees for height and underplant with characterful shrubs such as *Sarcococca*, hydrangeas, or *Nandina domestica*. Low-growing hellebores, ferns, and other perennials can be woven in between, with flashes of seasonal excitement from small bulbs.

Make the most of form by using the strongest and most interesting species available. Deploy bamboo, pampas grass, or clear-stemmed trees for their towering verticals, and roll out great dollops of evergreen grasses such as *Festuca glauca* or cultivars of *Carex*. Plant shrubs and trees with branches that fan outward like the tail of a peacock and elevate the acceptable to exciting with judicious pruning, raising the canopy of shrubs (see p.69) and creating intriguingly bony espaliers.

Go large with small plants. More is definitely more in the winter garden, so don't be too moderate when it comes to shopping for plants or dividing them. A few bulbs here and there can look indecisive and spotty, so go for a full-on swathe. Even if the plants are not all the same, a consistent theme of yellow, blue, white, or purple will look harmonious.

Create a framework with evergreens. These plants hold a garden together and provide year-round structure and coherence. They can also be used to lead the eye or bounce the onlooker from place to place around the garden. Clipped box and yew are traditional, but *Ilex crenata*, pyracantha, and osmanthus can all be sculpted to taste (see p.126).

Celebrate the flowers, as they can be few and far between. With the honorable exception of certain rhododendrons such as *R. dauricum* 'Mid-winter' or *R.* 'Christmas Cheer' and a number of tough camellias, there are not many truly large winter flowers. But those that there are make up for small size with delicacy and perfume. *Hamamelis* has charming lemon-zest tassels and *Lonicera fragrantissima* may be scruffy but it smells divine (see pp.40–41). Hellebores and snowdrops are addictive, and clever use of small bulbs (see pp.140–143) will keep the garden in bloom throughout the season.

Clockwise from top left A striking carpet of *Bergenia* 'Wintermärchen' under the red stems of *Cornus alba* 'Sibirica.' *Prunus × subhirtella* 'Autumnalis Rosea' is indescribably dainty. The dark foliage of *Ophiopogon planiscapus* 'Nigrescens' sets off snowdrops nicely.

A Different Perspective

The craft of designing gardens is about creating an experience. It is about applying maths and geometry and working with lines of sight and vanishing points, together with manipulating space and telling stories.

Garden design is a three-dimensional realization of ideas, and in an abstract sort of way, it is not dissimilar to playing a computer game or creating a simulation. You start off with a grid or a framework, upon which is imposed a view and a landscape that is all about angles and perspectives. Interesting features move past as you travel, and if you look to the left or the right, things appear. There are even rewards and treasures to collect, or at least note and admire, as you proceed through the room or along the course.

Whether the creation is in the mind's eye, sketched on paper, or drawn up using a design package, it is about the journey through an immersive space: leading the way through the landscape, with paths and marching topiary forms, and by using repeated hits of yellow disappearing into the distance like a trail of vegetal breadcrumbs. There are promises of something good around the next corner, hidden things and surprises, and a storyline; a system of suspense and reward compounded by an associated hit of dopamine, the brain's pleasure chemical.

Unlike louche, untidy summer, winter gets close to the bones of existence. It is now possible to see the framework, to paint three-dimensional pictures with plants and time, and take the ideas to the next level. Just like a computer game, it is about creating a magical world, but in reality.

Right A gentle meander slows your feet and demands that you take in the highlights of the season, while sentinel yews add emphasis and rhythm.

Creating a Habitable Space

Gardens are for living in, as well as tending and looking at, but when winter arrives, remaining warm and dry is a significant part of having a good time. Wrapping up well will help, while an outdoor refuge of some kind can also considerably extend the amount of use the garden gets.

While paving and infrastructure are important to the functionality of the garden (see pp.102–115), dry feet and a nice view do not necessarily equal a relaxing experience, especially when the weather is chilly or damp, or when there is a brisk north wind blowing. When thinking about how the garden may be used in winter, therefore, it is important to include a covered area in which to sit or, at the very least, some means of keeping dry.

On a modest scale, the way that windows and doors connect the house to the garden is part of this, and in many moderate plots, a pretty, oversized umbrella and a comfortable seat in the greenhouse or conservatory will suffice on a damp day. But where there is space, a permanent structure that enables time to be spent outdoors in inclement weather can be an asset.

Standard garden canopies and shade sails are not usually up to the buffeting of winter, so the hardy, every-season gardener must improvise. Effective and fairly low-key solutions include building an open-fronted shelter or a sturdy gazebo. Existing summerhouses, sheds, or garages can also be adapted and beautified in keeping with a glamorous winter lifestyle. Incorporating all-weather seating and outdoor furniture keeps things flexible, as does storage for a selection of warm wraps and rugs.

Adding lights to a dry and sheltered area can make it usable both day and night. This might be a string of fairy lights that can be switched on from the house, or lanterns hanging from the rafters and a table set for a bravely alfresco supper, decorated with flowers and foliage (see pp.176–179).

Storage solutions

Every garden needs storage, but what form this takes can vary. Some items can be left outside for months or parked behind a bush where they won't be seen, but when the rain is pouring down and the plant is suddenly see-through, a proper place to put things assumes greater priority.

The accoutrements of a civilized cold-climate lifestyle, such as blankets and cushions, can live comfortably in a trunk in the shed or summerhouse, or be stowed into built-in benches. But horticultural practicalities rule and, where possible, a garden should include a dedicated screened area that is invisible all year round, but represents both the engine room and hold.

This will solve the conundrum of where to keep unused pots and bags of potting mix, and if there is space for a discreetly hidden shed so much the better, as this will provide dry storage for tools and machinery, together with pots of overwintering dahlias and drying garlic, beans, and flowers.

Left This modern oak summerhouse is sculptural all year round, and even in winter it offers a dry spot in which to sit and contemplate the tall grasses and shimmering reflections.

Connecting *the* Seasons

The movement of the earth around the sun is a constant, bringing with it equinoxes and solstices, together with the fluctuations in temperature and light that affect the way that the garden grows.

Yet, for all our romantic attachment to summer love and sunshine, the seasons gently turn too, each running into the one that follows. Time leaving fall in its wake; the current presence of winter, and the near future of spring.

Some plants are known for their seasonal performance. Daffodils and cherry blossom are a certain sign of spring, while the bright leaves of acers and liquidambar signify fall at its best. But while these highlights are dramatic, they can disadvantage the plant in question, eclipsing other qualities such as good form and bark texture, and meaning they are less likely to be chosen for year-round interest.

To make a garden that is truly strong, it is important to select plants with multiple good qualities and several seasons of interest: the acers that add structure and zingy young foliage to their spectacular fall color, and amelanchiers that also produce berries to feed the birds; crab apples that persist until early spring, to be followed by pretty blossom, and camellias that provide evergreen foliage and bold early flowers.

Also valuable are long-lasting herbaceous plants and grasses with attractive seedheads, *Rosa moyesii* or *R. glauca* with their handsome rosehips, and the ornamental thorns of *R.* x *cantabrigiensis*, *Berberis*, *Zanthoxylum*, and some species of *Euphorbia*, all of which have interest that extends into winter.

Right Ghostly trees and softly faded grasses are still beautiful in the hazy light of a winter dawn.

Discovering Hidden Gems

The big reveal that comes with the arrival of winter can lay bare a garden's failings, but it can also demonstrate that it has far more to offer than originally thought in terms of planting space and opportunity.

Obscured by leaves no longer, large patches of empty ground present themselves to the opportunistic horticulturist. An exciting prospect and new space to fill, this is ideal for the plants that are adapted to take advantage of the sudden rush of light—the small bulbs such as snowdrops, *Eranthis hyemalis*, and cyclamen, and hellebores and foliage plants such as arums, together with pulmonaria, heucheras, and bergenias, as long as the shade at other times is not too deep. A scattering of primroses will then take the area through to spring.

Out of season, mature shrubs and trees reinvent themselves in new ways. Large hydrangeas, rhododendrons, and lilacs add permanent structure and height to an herbaceous border or woodland edge, where their attractive bark or persistent flowerheads make a pleasant contribution.

Some plants work the other way round: they are wonderful in winter and its shoulder seasons, but need only be charming in a busy summer garden. These include early-flowering cherries such as *Prunus* × *subhirtella* 'Autumnalis,' *Stachyurus praecox*, *Euonymus* species such as *E. alatus*, and *Clerodendrum trichotomum* with its fall berries, along with a number of core winter-flowering shrubs including *Hamamelis*, *Corylopsis*, *Lonicera fragrantissima*, and *Edgeworthia chrysantha*.

Using every inch of the garden, and adding exciting details specifically for this time of year is an investment for the future: for years to come, pottering about on a dry morning and standing in quiet contemplation will be enriched by a little time and attention spent on planting now.

Faux-frost effects

Plants with pale or variegated leaves can give the whimsical impression of frost when there is none. Gray- or white-tinted shrubs include cultivars of *Pittosporum tenuifolium*, *Euonymus fortunei* 'Silver Queen,' and *Fatsia japonica* 'Spider's Web,' while grassy options include *Ophiopogon planiscapus* 'Little Tabby,' *Carex* 'Silver Sceptre,' and *C. oshimensis* 'Everest.'

There is also a range of silver-gray or bluish options, such as lavender, *Convolvulus cneorum*, or *Santolina chamaecyparissus* 'Pretty Carol,' together with glaucous cypresses or *Festuca glauca* 'Elijah Blue.' For stems with a waxy bloom that evokes a touch of frost, try *Salix acutifolia* 'Blue Streak,' which likes a sunny, damp spot; shrubby *Rubus cockburnianus*, or *R. biflorus*.

Plant rescue

As the garden grows and develops, some plants will naturally do better than others, and arrangements that at first were working well can come unstuck. This is a particular issue when combining small plants with larger ones—as a small evergreen becomes less so, or a shrub starts to spread, snowdrops can be strangled and grow desperately lanky, and crocuses and early perennials may pine for the light.

It is therefore worth keeping an eye on how things are getting on and checking that all the plants are still playing nicely together. If partners have outgrown each other—and particularly if a plant is marauding to the degree that you have to delve underneath it to find its sorry neighbors—move the weaker party out of harm's way and put it somewhere where it can thrive.

Clockwise from top left Chinese lanterns (*Physalis* species) are easy to grow and their dramatic orange bells last for months. A great shrub for year-round interest, *Corylopsis glabrescens* bears fragrant flowers in late winter and early spring. The sterile bracts of *Hydrangea macrophylla* last for many months and are revitalized by a touch of frost. The ice-spangled thorns of *Berberis vulgaris*.

Packing *in* *the* Interest

When developing a garden with winter in mind, it is more important than ever to use the available space wisely. Walls and pergolas can still be clothed with climbers, and lawns freshly filled with bulbs. At the same time, managing both deciduous trees and evergreens so they have maximum structural impact, while leaving the surrounding area open and bright enough for other plants to thrive, is always a winning move.

While many plants are adapted to the dappled shade of deciduous woodland, it is much harder for anything to grow in deep shade, where dense foliage blocks out the sun and photosynthesis is impossible. Mature evergreens such as yew and holly are a particular challenge, as they shield the soil from both light and rain. As good hedging shrubs, however, these can simply be clipped back to the point where they don't impact on their surroundings too much, except in the sense of providing an attractive backdrop.

Top far left These two holly trees have been pruned as standard lollipops, which has enabled the ground beneath to be planted with a range of other interesting elements, including euphorbia, grasses, and *Physalis*.

Top left In late fall, the peeling bark of *Acer griseum* acts as a textural focal point against a tapestry of frosty seedheads, winter trees, and shrubs, all clinging onto their last few leaves.

Bottom left *Parthenocissus henryana* is more delicate and less thuggish than standard Virginia creeper, but it has the same glorious fall color and the emerging leaves are an attractive bronze hue.

Woody plants can be left to grow naturally, but a more exciting effect can be achieved by "legging them up," or raising the crown. This is a process which involves removing the lower twigs and branches to expose the structure of the supporting stems, in order to create a much lighter-feeling multistem shrub or small tree. It is particularly effective with mahonia and camellias, together with other heavy-leaved shrubs such as holly, bay, and photinia, which might otherwise dominate in a small garden.

In this way, the design gains a more elegant and sculptural element and the light and water that reaches the ground allows close underplanting with hellebores, evergreen grasses, and bulbs. It is also possible to use the same trick on deciduous plants to improve their form, to create a clear stem, or to create an airy umbrella effect. It is a good way of integrating crab apples, for example, and plants that are attractive but tend to be bulky, such as *Sambucus nigra* 'Laciniata.' As with evergreens, using the technique on deciduous shrubs helps them perform well in the context of their surrounding plants and opens up further opportunities.

Using vertical space

Training plants against a wall is a very effective way of maximizing growing room and structure while minimizing the footprint of the plant. Ivy and deciduous *Parthenocissus henriana* will happily scale a vertical surface with very little intervention, and training pyracantha, osmanthus, or cotoneaster flat against a wall, or as buttresses around a door, can generate a really striking feature.

Arches or obelisks will also provide vertical support for scrambling *Trachelospermum jasminoides* or *Clematis urophylla* 'Winter Beauty,' as well as having a structural role of their own.

Making a Winter Garden

Working
with Light

In the winter garden, light has many facets. It is not just the absence of darkness, but also a useful tool; a dramatic device and the bringer of health and hope. Inconsistent though it may be, it is an element without which life cannot survive.

The sheer fluidity of light in the garden is complex and intriguing, and to understand the way in which it affects any given place takes time. There must be a willingness to stand and stare, to watch how the angle of the sun relates to the aspect of the site. Noticing where the light falls and how it moves throughout the day; seeing where the rays slice through, creating patches of brightness and emphasizing the opposing shadow.

At night, light is a force to be harnessed, as our tool and our toy. Yet, by day, it rules us; we pursue it and hunger for it, keeping vigil as we await the first sign of growth, anticipating the flowers of hope that reassure us that the sun will return.

In this dullest of seasons, light is something to be embraced and cherished. The days are short and the landscape bare and muted, but light is energy and power, and however much or little there is, it should be used thoughtfully, and in the most effective way.

The Art of Illumination

Fluid and inconsistent, light is an element without boundaries. It obeys the laws of physics and underpins biology, yet it is also the paint and the poetry that gilds the winter garden with glory.

While a garden is, at its core, a collection of plants, it is also a living, breathing system; a composition or a work of art that is more than the sum of its parts, and light plays an important part in this.

The nuanced requirement that plants have for this form of energy underpins much of horticulture, yet it is at the same time a creative medium like no other; one that can be combined with the delicacy of the season, and used in a truly painterly fashion.

Capturing light is a challenge for artists. The French Impressionists excelled at it, filling their paintings with dappled shade, scintillating seascapes, or the play of the sun on meadows, each glimmer captured with a fleck of paint. Photographers, meanwhile, sate themselves with the brief opalescence of dawn and the liminal perfection of day meeting night.

In a garden, however, light is not captured and held at its most exquisite moment, nor is it echoed on canvas for posterity. Applying the principles of design to a living system requires an acknowledgment that it will inevitably change with the tide of the seasons, and that it will evolve and grow. Within this, light is a dynamic and mercurial element, one that moves infinitely faster than the earthbound vegetation. A brilliant sky can be obliterated by a sudden downpour and sunshine momentarily dimmed by a wisp of high cloud before bouncing back. And, at a chemical level, the plants react. This is part of the deep pulse of existence: an ongoing cycle of light and shade, the cellular response of living things to solar fluctuation. It is a constant balancing act that creates an excitement and vibrancy in the garden, and everything that lives within it.

Working with light requires a synergy between observation, gut feeling, and skill, but mastery of its whims and moods elevates the discipline of gardening like nothing else can.

Left The contrast of light and shade in the winter garden brings movement, texture, and a renewed sense of energy.

The Facts of Light

Without the sunlight that beams through the universe and suffuses our planet with warmth and energy, life as we know it would not exist. The fact that it does is a small miracle even so, and, in turn, we owe our own existence to plants.

When a photon hits chlorophyll, a cascade of reactions combines carbon dioxide with water. This creates sugars to fuel the plant's growth, and releases oxygen as a by-product. But, at a molecular level, plants are conducting a constant cost-benefit analysis. In winter, there is less light and the temperatures fall. This affects the rate of both photosynthesis and respiration, so growth slows and many plants drop their leaves and become functionally dormant.

Human beings, too, respond to light in various ways. We usually sleep when it is dark and become more active during the day, and research has shown that bright light can intensify emotion and improve low mood. The quality of our vision is also linked to the amount of light that is available. Our eyes contain two different types of photoreceptors, called rods and cones. Rods work well in dim light, but they see only shades of gray and they also have low spatial acuity, which is why everything looks grey and fuzzy at dusk. Cones allow us to see color and they operate best in bright conditions; there are fewer of them but, unlike rods, they are densely packed for high spatial acuity, and this allows us to see detail.

On a fundamental level, this is why the quality of light is so important to our enjoyment of winter gardens. In dim light we don't see color as well— or at all—and the scenery around appears dull. Not all colors are equal, either. Although white and pale colors remain visible at low light levels, another process skews perception away from the red end of the spectrum, meaning that reds will seem darker.

The relationship between light and design

The impact of light on winter garden design is considerable. On a basic level, the design has to take into account that plants have evolved according to the availability of light; dormant plants don't need light while growing ones do, and some need more than others. Light also affects aesthetics, and the positioning of plants should ideally take into account visual adaptations, for the best effect.

With this in mind, placing red and orange cornus in the brightest spot and backlighting the peeling copper bark of *Prunus serrula* makes perfect sense, since the colors are literally easier to see. The pale appeal of classic winter garden subjects such as birches and snowdrops also makes sense, in the context of their enhanced visibility. Arranging for the setting sun to strike seedheads and highlight their delicate tracery is likewise obvious: in a shady position the details would be less clear.

It is an elegant balance of physics, biology, and biochemistry overlaid with the essence of humanity and psychology, where the study of light, gardens, and our own ability to perceive comes together in a triumph of evolution. And, as a concept, it is beautiful.

Top right With bold, repeating verticals against a delicate silver filigree of stems and seedheads, a clear, frosty morning shows this garden at its best. Here, *Malus × robusta* 'Red Sentinal' is underplanted with *Spirea* 'Arguta'.

Bottom right Due to the way our eyes have evolved, we are better able to see reds and oranges in bright conditions, so colored stems, such as this *Cornus sanguinea* 'Midwinter Fire', are most vibrant when grown in a sunny spot.

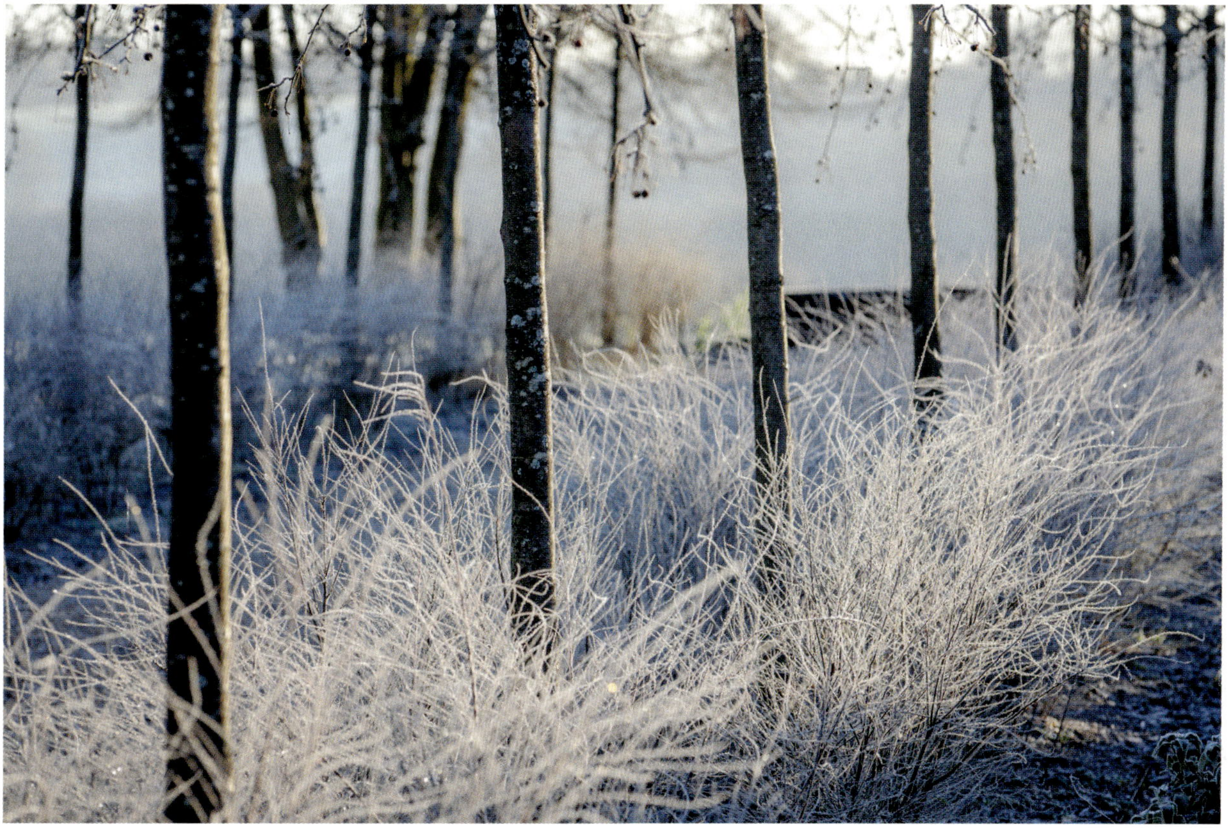

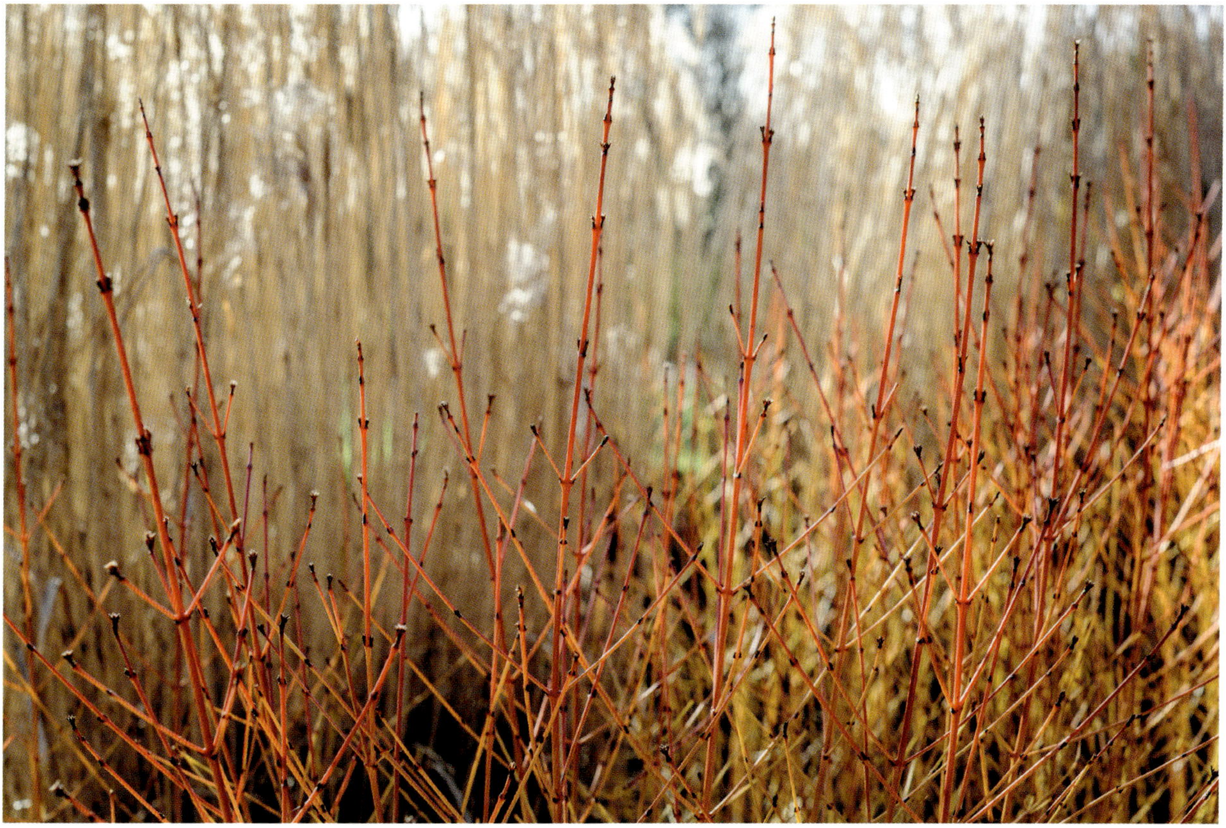

Making *the* Most of *the* Sun

For half of the year, the garden is suffused with light that chases shadows away and fills the space with warmth and abundance. Ever-present sunshine—or daylight, at least— is the expected way of things. But near the equinox, a sudden chill comes. The light loses some of its strength as the days shorten and one season tips into another. A rising tide of shadow consumes parts of the garden, and for the things that grow there, the environment changes significantly. It is a point of reckoning.

As the light retreats, it is important to understand how the garden deals with its absence. Notice what is cast into shade and for how long those shadows last. A shaft of light for an hour or two in the morning, and shade the rest of the time, is one thing; the sun setting behind nearby buildings in late fall and not rising high enough to strike even the tallest plants in the garden until early spring is something very different.

Top far left Backlit trees cast long shadows over the clean canvas of a frosty lawn; their inverted shapes sprawl across the ground, adding a new dimension to the landscape.

Top left A sunny day will emphasize shapes and textures in the garden, and bring the more muted colors to life.

Bottom left Breaks in a backdrop of trees, or judiciously cut into a hedge, act like windows, not just permitting a view out of the garden, but enabling sunlight to flood through the arrow-slit to illuminate carefully positioned features at key times of day.

Learning the space is invaluable, whether through keeping a diary or just by feel. Take all the time that is needed to think about where the sun comes from and what it illuminates at different times of day. Observe the refreshing, brightening effects of a frosty morning or a sprinkle of snow.

Shade requires a different garden, but it does not have to be a bad garden and there are plenty of plants that will thrive, regardless (see pp.80–81). It can be a place of cool and quiet, a meditation on the versatility of green and characterized by plants such as mosses, ferns, and other woodlanders that enjoy a dimly lit location, together with acers and a selection of other shrubs that eschew bright light and summer heat.

A shady corner can also be punctuated by interest other than plants—a piece of sculpture, for example, might evoke classical decay or something boldly modern; thoughtfully placed rocks or lanterns can be used to create the impression of Japanese style.

Where the light makes its presence felt, the interplay between buildings, structures, and trees can be an asset. Use shadows, and the shafts of light between them, as a dramatic device to spotlight a key feature, or simply to warm up a well-placed bench in time for morning coffee.

Place plants carefully to take full advantage of the effects of the sun, siting trees to catch the early tendrils of dawn in their branches, and putting bleached seedheads and bright stems where they will be lit up by the golden evening rays.

Clever planning enhances the best qualities of a plant, revealing its translucence and boosting its brilliance. Backlit by the sun, stems and faded perennials are thrown into black-and-gold relief; birches become a transient, twiggy filigree and oaks anchor an opal dusk with a sense of strength and permanence. Low light turns crocuses into painted glass and adds a glossy brilliance to snowdrops, while in the fresh silence of a frosted dawn, spun-sugar grasses and gaunt shrubs become fine and luminous.

The Challenge of Shade

In the garden, the presence or absence of light is a fluid situation, but total shade, whether transient or not, needs to be accommodated. It is a fact of life. It is pointless to ignore it, and to do so risks sizable parts of the garden becoming underutilized—not just in winter, but potentially in summer too.

Assessing your shade

While it is important to capitalize on the available sunshine, it is also important to acknowledge the areas that see none: the corners that now spend most of their day in shadow, and the small gardens that are tucked behind houses and high evergreen hedges, creating places that may not see direct light for several months.

Yet the world turns and the seasons wax and wane. It is not a binary case of sunny or shady, one or the other, and the quantity of light is influenced by the time of year and the angle of the sun, together with the proximity of buildings and of surrounding plants.

While buildings are static, the plants are a variable—and when they are a substantial, woody variable that gets bigger every year, their effect can be considerable. The nature of the plant is also significant. Evergreens cast a reasonably consistent amount of shade, while deciduous shrubs provide shady spots when in leaf, but let the sun through in winter. Even the quality of the foliage matters; the difference in light penetration through an airy birch and a hefty sycamore is marked.

When approaching a site in midwinter, there are some key questions to ask. Is it shady now? Is it always shady? Think about when it gets the light, which months in the year and which hours in the day; whether that shade is created by the dappling of a tree or by a solid building. All these things will affect what will grow.

Choosing plants for a shady spot

However good a design is on paper, growing conditions will always triumph and a statement plant that sings in the sunshine will be wasted in the shade. Consider the space, therefore, from the perspective of horticulture; find out what will thrive, both in winter and in summer, and work with the unique charms of the plants that will be most happy.

Cyclamen and snowdrops usually do well in the dappled shade of deciduous shrubs, and even in the temporary heavy shade of houses, as long as the sun comes round as the season moves on so the plant can photosynthesize. Avoid sun-loving crocuses and colchicums, however, as the flowers will etiolate and flop in damp shade, and they are vulnerable to slugs.

Not all evergreens are equal, either. Cedar and juniper don't like shade at all, while cryptomeria and pittosporum are more tolerant. Evergreen ferns love a cool and somewhat dank spot, camellias are fairly easygoing, and *Hydrangea macrophylla* may not be in leaf, but it holds on to its flower heads for months.

A border that has summer sun but winter shade, meanwhile, is fine for herbaceous perennials and deciduous shrubs. They will be dormant at the point the lights go out, and as they start to grow, the sun returns. In seasonally shady spots, some of the less fussy fruit trees and roses are even worth a try, forgiving the location while they are not growing, and providing useful winter structure.

As ever, it is important to put the right plant in the right place. For both rosebushes and crocuses, a strong preference for light in growth is often a virtual irrelevance when dormant—whether this state is in winter or summer. Seasonality can be key, especially with plants that time their activity to coincide with leafless trees, while others simply embrace year-round shade and carry on living their lives.

Clockwise from top left Elegant *Camellia* 'Shiro-wabisuke' flowers in late winter. Semievergreen *Dryopteris wallichiana* copes well with part-shade. Snowdrops bring light to the gloom. Given lighter summer conditions, *Pittosporum tenuifolium* 'Tom Thumb' tolerates winter in the shadows.

The Power of Reflection

With the leaves gone and the landscape open to the sky, reflections come naturally to the winter season. In nature, a glassy lake or meandering waterway will reflect soaring mountains or pencil-sharp pine trees, inverting the world and adding a new and spectacular dimension. But the essence of reflection, and its own particular magic, can be harnessed to work in smaller spaces, too.

Installing a simple garden water feature can be highly effective. A raised pond or formal canal, sited to make the most of the sky to the west, will dramatically catch the last rays of the setting sun. Almost any container that is wider than it is deep, such as a tank or bathtub, can be repurposed to form a mirror pool and positioned under arching branches or next to a bold or colorful planting scheme, where still water or thin ice can echo reality and double the impact.

Where water is impractical, an alternative is to create a reflection with a metallic sculpture or an old mirror, hung on a wall and half-hidden by leaves for an artistic effect. Both will bounce the light around the space, adding movement and tricking the eye into thinking that there is something else that might be seen, while a discreet location and foggy quality will minimize the risk to birds, which may mistake a very reflective surface for a flight path or their own reflection for a rival.

In courtyard gardens, meanwhile, adding a coat of paint to a south- or west-facing wall or fence will gently direct some of the incident light into the darker corners, and suffuse the space with a soft brightness.

Right Striking reflections in the lake at RHS Wisley, England, UK.

Using Garden Lighting

When long hours of darkness limit outdoor activities, adding lights to the garden can be an efficient and attractive solution. As well as making the space secure and safe to use, cleverly planned illumination can also transform an inky view from the windows into an inviting and magical nocturnal world.

Garden lighting has two main functions. There is the practical task of illuminating paths and steps, and it can also be used for the aesthetic purpose of highlighting plants and structures, or creating attractive features or installations. In both these cases, there is a huge amount of variation and choice.

Functional does not necessarily equal unattractive, and there is a strong argument for new installations to be visually pleasing, no matter how prosaic their purpose. Marking paths with solar lanterns, fitting recessed tread lights onto steps, and adding lights that simultaneously pick out shrubs and indicate which route to take—all achieve the same result as a motion-triggered floodlight, but with a completely different feel and experience: one that is pleasant and welcoming, rather than blasting away night vision and creating deep and lurking shadows.

Aesthetic lighting, meanwhile, can be used in a number of ways, from versatile spike lights, which are pushed into the ground and then directed at the feature of interest, to an enormous range of uplighters, downlighters, and LED strips.

For atmosphere, it is hard to beat a natural flame, and candle lanterns or firepits are perfect for a cosy evening snuggled under a blanket. Globe lights and solar lanterns make attractive features, while Moroccan-style lamps cast characterful shadows. Even something as simple as fairy lights or swags of festoon lights can evoke a festival vibe while also illuminating the way.

Light pollution

It may be tempting to brighten short days and dark nights with a million electric stars, but lights should be turned off when not needed. Saving energy and reducing the electricity bill makes sense, but it is also important to limit light pollution as much as possible.

Life has evolved circadian rhythms that dictate behavior, sleep, and hormonal activity over a 24-hour cycle, with measurable periods of day and night. Disruption from artificial light affects not just humans and plants, but a wide range of other organisms, including mammals, birds, amphibians, reptiles, insects, and even bacteria, and nocturnal species are disproportionately impacted.

The examples are legion, from turtle hatchlings that head for the twinkling lights of a coastal town rather than toward the sparkling sea, to the behavioral changes of garden insects. Moths lured to lamps are easy prey for opportunistic predators or, in their fascination, they may become exhausted and die. The creatures that prefer the dark remain in hungry hiding, meanwhile, and when night-flying pollinators are disorientated by artificial light, plants suffer too.

Lights are vital for safety, but they don't have to be on permanently. Installations should accord with dark-sky principles, and when the lights are just for decoration, an automatic timer that switches them off at bedtime will benefit the neighbors of all species.

Top right Gentle illumination of key plants and focal points brings a painterly quality to the night garden.

Bottom right Lighting can add real texture and emphasis to a garden design, enhancing features and creating a sense of journey.

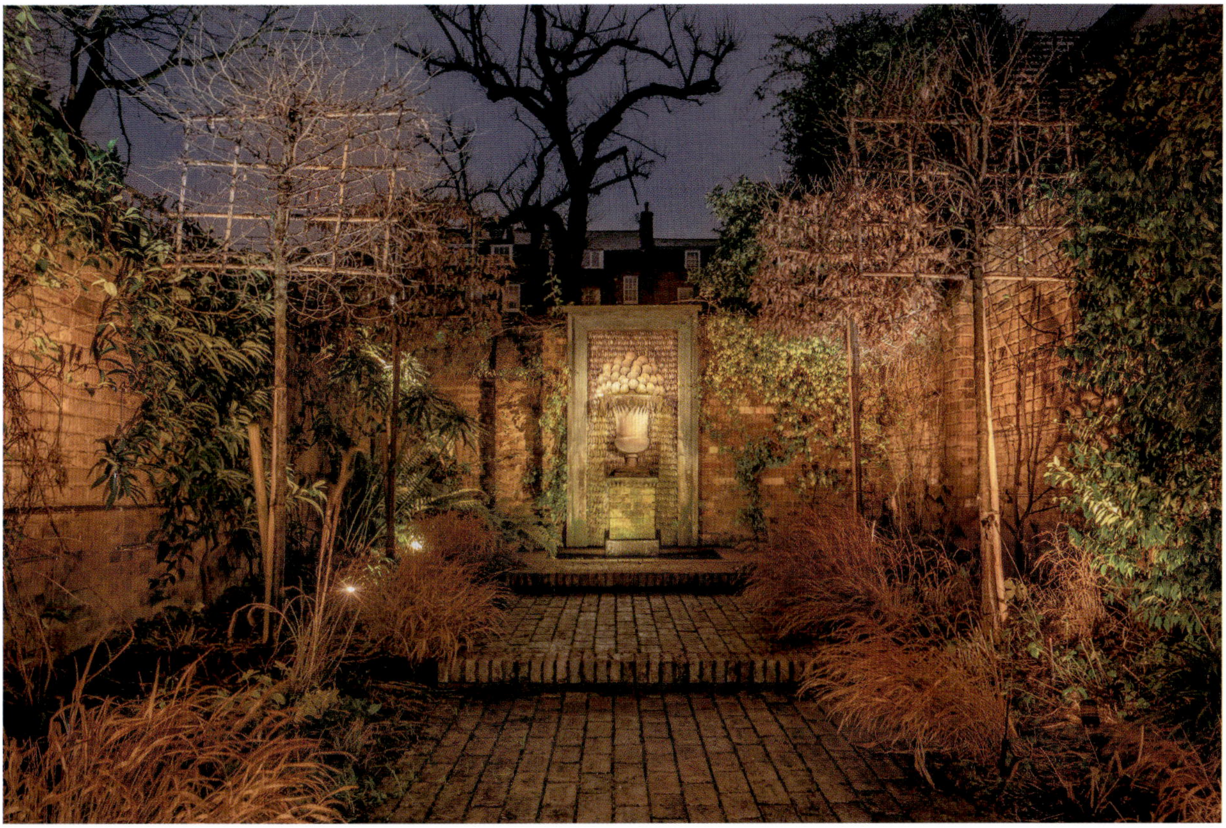

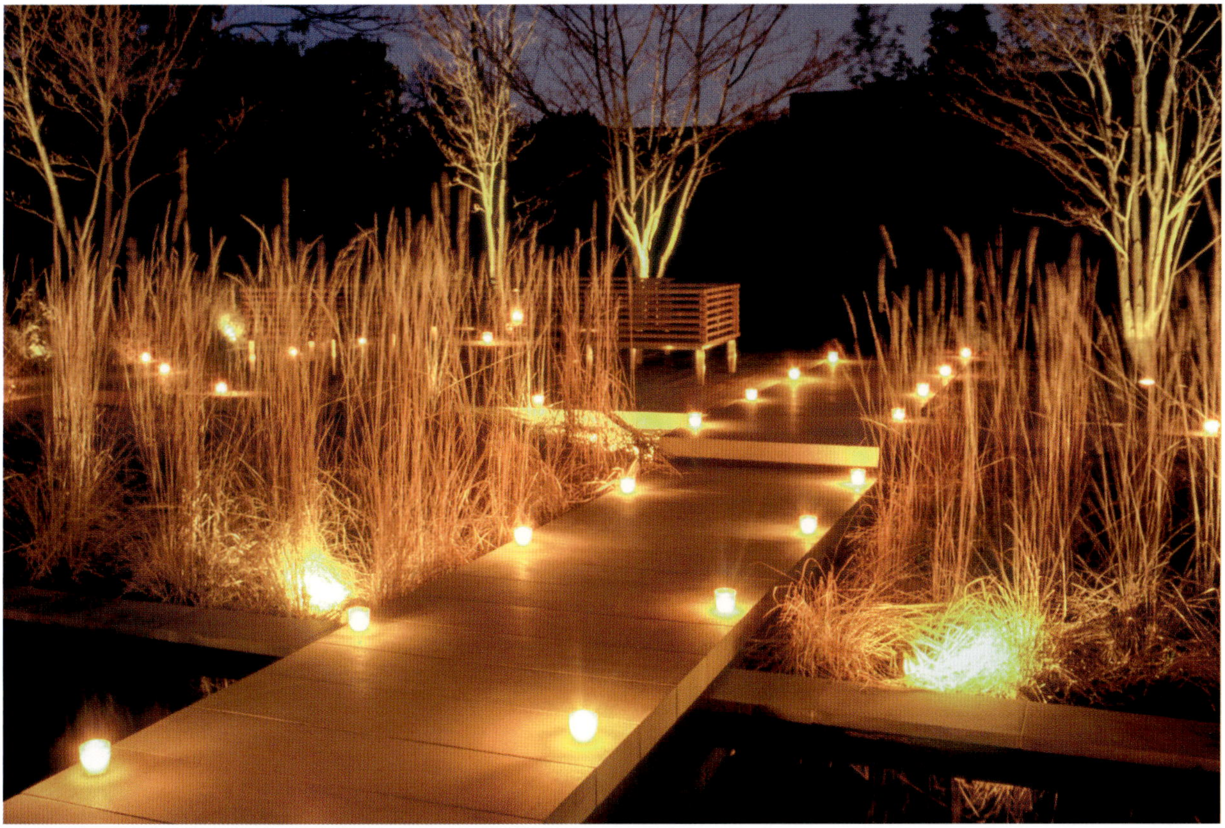

Winter Containers

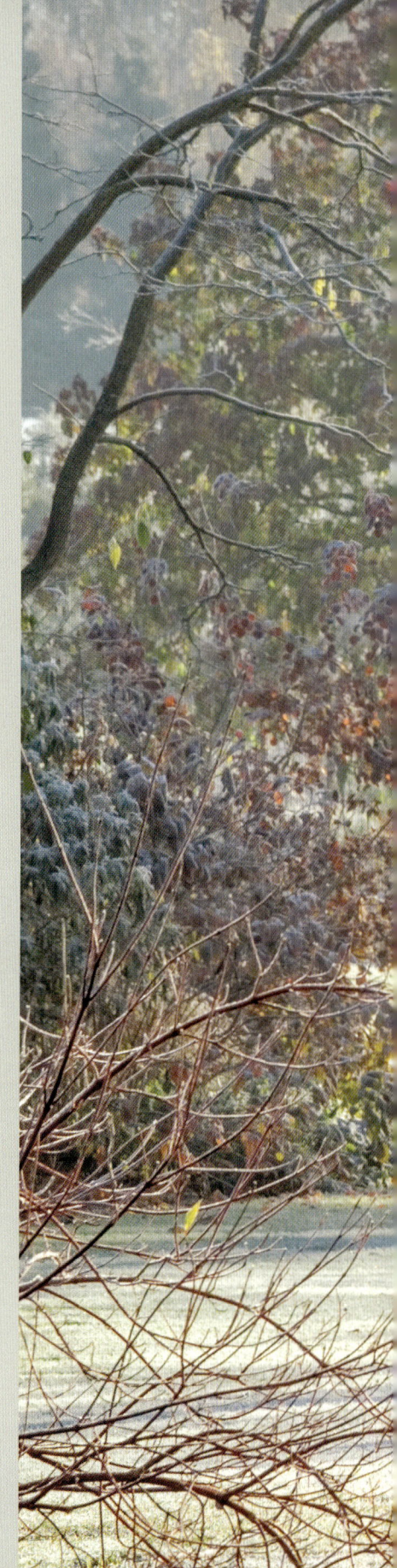

In the winter garden, containers can be a lifesaver. Nothing adds a shot of color or lifts a lackluster vista as quickly as a tub of fresh foliage, some bright bulbs, or a splash of cheerful bedding, while less competition from its surrounding landscape means that the pot itself can be a statement of taste and style, too.

A container can be seen, essentially, as a garden in microcosm: a portable and affordable freestanding planting scheme that oozes attitude as a standalone feature, or that can be fine-tuned to match or harmonize with neighboring pots. What's more, in a season known for its horticultural challenges, containers offer the gardener a solution that is supremely adaptable.

Elsewhere, the garden may be good, bad, or indifferent—or may just need a few seasons for key elements to grow large enough to hold their own—but a few planted tubs can add an instant hit of excitement and an opportunity to tweak the view as fall turns into winter, and winter heralds spring.

Containers *in* Context

The beauty of planted pots is that they deliver regardless of their surroundings. With fewer garden plants in full show-off mode, container arrangements stand out in an uncrowded field, offering the opportunity to celebrate the season, to experiment and to be daring, without the challenges of long-term commitment.

A versatile part of the gardening toolbox, containers excel no matter what is going on in the rest of the space, and even if there is no proper garden at all. The choice of pot, however, is fundamental to the overall impression. This is the base point and the anchor; the element that delineates the extent of the container as a feature and does the visual boundary work for the plants that it contains. What works best, therefore, depends very much on the style of the garden, the tastes of its owner, and the size of the planting scheme in question.

With the pot in place, it is time for some creative horticulture. Unusual specimens whet the appetite; fascinating berries demand attention, and evergreens of all kinds can hold their own for months. Some combinations are subtle and stalwart, such as heucheras, *Ophiopogon*, snowdrops, and hellebores, or harmonious arrangements of two-tone pittosporum underplanted with pink, purple, and silver. However, this subtlety is entirely optional and plants can have a foot in both camps.

Left In a small space, clustered pots filled with flowers and foliage will provide a long-lasting display, while the black backdrop adds an additional element of theater.

Phormiums are vibrant in color and energetic in form; *Cornus sanguinea* 'Midwinter Fire' can look like a static explosion of light; while lashings of screaming pink cyclamen and punky *Carex oshimensis* 'Evergold' blast color at the onlooker, blowing gloom and cobwebs away, brightening the view and providing a welcome distraction from any short-term failings elsewhere.

Taking a longer view, container gardening can be used as a trial ground for glories to come. Temporary combinations in a pot give the gardener a chance to get to know each subject before committing it to open soil. Since all plants and gardens are different, and nailing winter performance is quite a specialist skill, it pays to become familiar with the available assets and consider the potential of each one.

Containers permit rapid change and forward planning, and represent a gardening element that can easily be altered, rearranged, or moved around on a whim. Clustered to form a tableau, and with new arrivals substituted as the season progresses, the container gardening concept can be scaled up or down as required, and precisely targeted to lift the spirits with punchy, upbeat planting just where it is most needed.

The Science and Practice of Winter Containers

To get the best results, it is important to understand how variables such as weather and scale affect planted pots, particularly if they are to be a permanent feature.

Water

Some moisture is necessary for growth, but how much will depend substantially on the type of plant and how actively it is growing. Vigorously shooting bulbs need more water—and food—than a deciduous small tree, where too much of either can cause problems.

Drainage

Soggy, stagnant potting mix can be fatal, so good drainage is essential. Make sure there are a number of good-sized holes in the bottom of the pot. Remove trays of water in winter, and watch out for overflowing gutters dousing nearby plants in a downpour. When rainfall is protracted, move containers into the lee of a building, or temporarily under cover.

Using a growing medium that drains really well can make it possible to push the limits of climate, so consider adding some coarse sand or vermiculite to the potting mix at the outset. Many more tender plants will withstand a few degrees of frost, and may overwinter outside as long as their roots are not languishing in cold, wet conditions. With sharp drainage and the shelter of a wall, even containers of succulents such as aeoniums or echeveria can often stay outside until spring.

Cold

All plants are more vulnerable in containers than they are in the ground, and in periods of cold weather it is important that pots don't freeze solid. Ice forming within the tissues can kill not just tender plants, but trees, vegetables such as broad beans, and even hardy bulbs.

Increase thermal mass by clustering pots together, and put the most vulnerable subjects in the middle. If there is a short, sharp chill, wrap the whole lot in one go using an old blanket, burlap, or similar. If the cold is persistent or temperatures are set to drop below 23°F (−5°C), move the containers into an outbuilding or insulate them thoroughly with straw, cardboard, or reused bubble wrap.

Potting mix

In containers, the growing medium needs to provide a steady supply of water and nutrients, and it must be well drained. In permanent pots, a loam- or soil-based potting mix is often used as it has a good, complex structure with plenty of air pockets to absorb and release the water. It also tends to be heavier than ordinary potting mix, which increases stability.

Soil-less "multipurpose" potting mixes are high in organic matter that will continue to decompose over time; the size of both particles and air spaces becoming gradually smaller. Eventually, the medium no longer drains properly, so the container may become waterlogged and the plant will struggle. In all potting mixes, nutrients will diminish too, so keep an eye on progress and repot as necessary.

All composts should be peat-free; price is generally a good indicator of quality.

Top right A substantial container filled with a good-quality, free-draining growing medium will be resilient in the face of winter weather.

Top far right Contrasting leaf sizes and ornamental berries add extra interest.

Below right With care, food, and water, a display of potted hellebores can be brought back year after year.

Size

In terms of temperature, nutrients, and water, a big pot provides a more stable environment than a smaller one. The larger the volume of potting mix, the more food and water it contains and the longer it will take to become saturated or to freeze. Smaller pots tend to dry out quickly, through a combination of plants taking up the water and the effects of evaporation. Porous materials such as unglazed terracotta exacerbate this effect and plants can quickly suffer from drought conditions, even in winter.

Overall, it is about getting the balance right. A large pot is resistant to freezing, has a greater thermal mass and more potting mix to protect the roots, and holds more water and nutrients to feed the plants in active growth. A small pot may be prone to freezing, but this is less of an issue if there is little water to freeze. This is worth bearing in mind when overwintering succulents or half-hardy plants such as salvias or potted dahlia tubers—some cold is manageable but not when combined with wet, so keep them fairly dry.

Tips for success

Plan well in advance by selecting plants and bulbs as the summer wears on, building on past successes and treating each container like a little design in its own right. A single type of bulb or a statement evergreen has a bold simplicity, or combine a tall focal point with something trailing or frothy and something flowering. Using several different elements will help to make the display last longer.

A well-composed mixed container, filled with good-quality potting mix (see p.90), should be able to last several seasons, but it will eventually become unbalanced. As plants grow, some get too large while others struggle with the competition, so when the display is no longer working, dismantle it, redeploy the rogue elements, and start afresh.

Left Recycling a weathered wooden box can look rustic or add vintage charm, while also potentially concealing cheap plastic pots.

Choosing a pot

There is a container to suit every taste and garden. The choice is extensive, but all planters should be frost-resistant and have decent-sized holes in the bottom to provide drainage.

Terracotta: natural clay pots weather beautifully, while glazed finishes introduce an element of color. Always check for frost resistance, as without it the pot may crack and fail.

Plastic: often inexpensive and increasingly recycled or recyclable, attractive plastic options include lead- or granite-effect planters, or slick contemporary styles.

Metal: available in a range of finishes, metal particularly suits a modern design. Root insulation, however, may be less good than other materials, particularly if the container is small.

Wood: versatile timber planters come in a range of styles and at various price points, and it is quite possible to make your own. Naturally frost-proof, wood can be painted or left untreated, but it will eventually rot.

Recycled or vintage containers: lending a distinct and personal twist to a garden, almost any vessel can serve a turn, but urns, baths, sinks, and washing tubs are ideal, as long as they have drainage holes.

In container gardening, proportion is key. Picking a vessel that is at least as high (or wide) as the plant is tall will give a better display—particularly with bulbs, which can have a substantial amount of root. A tiny pot can make even a small plant look top-heavy, but a weighty, sandy potting mix, or layer of pebbles in the bottom, can help stop it blowing over.

Root-prune containerized shrubs or small trees every few years and repot in fresh mix. This removes thick anchoring roots and encourages fine feeding roots, and helps to stop the plant from outgrowing its pot.

Container Inspiration

Rather than letting displays of potted plants fade away in winter, give them a new lease of life. Filled with flowers, foliage, bulbs, and berries, and allowed to take center stage, they will get better and more interesting as the season wears on.

A fresh array of planters is a quick way to energize a garden at any time of year, so when winter comes, refresh the scene in honor of the new season. Some stalwart arrangements will carry on regardless of the chill and can be put in pride of place, while others, such as saxifrages, sedums, or thrift may no longer be in bloom, but will still fill a gap attractively.

Planting up containers of bulbs is an obvious step in the right direction, but it pays to take a mid-fall nursery trip to get some ideas and stock up on plants that are fresh and new. Evergreen grasses and sedges often look great at this point, especially lovely, steely blue *Festuca glauca*, *Carex oshimensis* 'Evergold' and 'Everest', and *C. comans*, all of which are very hardworking and relatively inexpensive to buy. Bedding plants such as pansies will also put on a cheap and cheerful show.

As is the case elsewhere in the garden, foliage plants can do a lot of the heavy lifting in winter containers. Hellebores start into both leaf and flower at the right time, while heucheras are indispensable; *Ajuga* is a useful filler plant, and ivies are hardy and versatile too. Shrubby evergreens make a good central

specimen; small conifers, varied *Pittosporum*, and variegated hollies are a good addition to the ubiquitous skimmia and photinia. Open-mindedness and experimentation can yield good results.

It pays to think about the staging of the container display, and to group them where they can easily be seen from the window and enjoyed, even when bad weather limits access to the garden. As highlights come and go, the pots can be refreshed as often as is necessary to keep the display going all winter.

An informal table or bench will give containers a greater degree of height and prominence, or consider a dedicated display unit that is tiered like a theater. When it comes to the overall impression, the finishing touches are key. A little moss in the top of a tub of bulbs or a mulch of decorative pebbles gives an array an intentional look, as do adornments. A few carefully placed rocks will give an extra detail, or create a theme with ceramic or glass spheres, metal spirals, or anything else that appeals.

Adding seasonal interest to permanent containers

Some plants are so bold and striking that they look fabulous all year round and need no help to steal the show in winter, while others retain the bones of loveliness but benefit from a boost.

To stop clipped evergreens and fruit trees looking isolated, surround them with a supporting cast of cheerful cyclamen, crocuses, or pansies, and a splash of contrasting foliage. With large containers, it is usually possible to tuck infill planting into the existing potting mix; alternatively, nestle smaller pots into the top of the tub, around the base of the central plant.

When herbaceous plants such as hostas have died back, use the space for something else: an opportunistic handful of bulbs dropped into the top of the pot in fall, for example, provides a nice hit of color before the perennial sprouts anew.

Left Clustered together, containers can create a long-lasting display. Add interest by combining pots of high-impact foliage with others containing bedding, and build height with a larger arrangement. Here, a rose with ornamental hips combines beautifully with silver-green variegated foliage and the bright berries of *Gaultheria procumbens*.

Creating a Show

A successful display of pots might require more input than a permanently planted garden does, yet it is a fast, fluid, and creative way to get results, as clever containers inject color, drama, and life into even the smallest outdoor space.

The process of creating a really good container display is a marriage between both pots and plants. The container must have the essential qualities of proportion and aesthetic appeal, as well as providing for the plants' needs (see p.93). What is planted, meanwhile, must be strong enough to weather the season, work well as an arrangement, and also be sufficiently eye-catching.

When a planter has a strong shape, a bold design or a bright color, a simple, strong scheme is best—perhaps a single type of plant such as a bushy *Sarcococca* or topiarized *Ligustrum*, which might then be densely underplanted with grasses such as *Acorus gramineus* 'Ogon,' heathers, or small-leaved ivies. But where all the attention is on the plants, a more modest and retiring choice of container is called for. Blocks of color can be extremely effective, particularly in winter when there is little of it around; this can be achieved with massed bulbs such as muscari or the bright foliage of *Nandina domestica*. The shape of the plants is also a factor: the spiky form of a grass or phormium will create a different effect to the relaxed charm of hellebores, and the two elements may not sit comfortably together, while the junction between the container and its contents can be softened by a plant that cascades and flops.

Taking containers to new heights

At any time of year, it is worth looking beyond standard pots, tubs, and window boxes, to get the most out of all available space. In winter, hanging baskets can be filled with bedding pansies and primroses, ivies, grasses, and other evergreens, and jazzed up further still with pots of pregrown bulbs. Sunk into the display, pot and all, these can be easily removed and replaced by later specimens. A sequence of, say, *Cyclamen hederifolium*, Tête-à-tête narcissi, and *Muscari* will lend ongoing interest to a container without having to dismantle the whole ensemble.

Fixing planters to a wall, meanwhile, enables vertical gardening to continue into the colder months. Filling them with a variety of evergreen plants and trailing foliage will give a display that is lush and colorful whatever the weather. To keep the highlights rolling as the season progresses, it is generally best to use the wall-mounted planter as a holder and keep the plant or arrangement in an inner pot, which can then be removed when it needs care. Ensure there is plenty of drainage, and watch for overwatering.

Dispensing with conventional containers entirely, kokedama is the Japanese technique of wrapping the roots of bulbs or small plants in moss. The trick is to use a fairly small plant: an ivy or fern is ideal, or around a dozen bulbs of a smaller species such as *Galanthus* or *Chionodoxa* (these preferably should be pot-grown in advance). The roots or bulbs, still in their soil, are wrapped in a blanket of moist, live, sustainably harvested moss and secured using black thread or gardening twine.

For drama, the finished kokedama can be suspended as a group from a tree or a pergola, or they can be placed on a windowsill or table. Water as necessary, either using a spray mister or by immersing the moss ball in a bucket for a short period, and plant out the focal specimen when the arrangement starts to look tired.

Clockwise from top left A small metal bowl is filled with winter aconites, *Eranthis hyemalis*, and topped with moss. Planted with hellebores, *Leucothoe* 'Zeblid' and *Vinca minor* 'Imagine,' this dolly tub makes a sizable and striking container. All sorts of plants and small bulbs can be used in kokedama, and *Muscari botryoides* 'Alba' is an elegant choice. Loose, green-tinged hellebores team well with hoop-petticoat *Narcissus* 'Spoirot' and *Pulmonaria* 'Silver Bouquet,' while flowing *Carex oshimensis* 'Eversheen' follows the rounded shape of the terracotta bowl.

Using Bulbs *in* Containers

Potted up in fall and flowering merrily from winter to spring, bulbs are one of the easiest and most effective ways to create a joyous splash of color, either by themselves or with foliage plants and other seasonal flowers.

Once a container is planted with mature and healthy bulbs (see p.93), a few simple things will influence their progress. One is water; too little will delay the flowers or inhibit them entirely, too much and the bulbs will rot. The other is food; if flowers are to return the following year, the bulbs need to be fed.

Bulbs work well in company, and adding evergreen plants such as ferns or ivies will provide color before the flowers arrive, while cutting a few interesting or colored stems (see p.193) and pushing them into the top of the pot gives height and adds interest before the bulbs even emerge.

It is currently popular to layer bulbs, stacking different varieties one on top of the other, according to their ideal planting depth. The idea is that these emerge in succession to extend the pot's interest. In practice, however, it is not particularly effective because the bulbs grow at different rates, which means that early-flowering varieties are obscured by emerging leaves and later ones are marred by dying foliage.

How to plant a pot with bulbs

Choose a pot that is big enough and tall enough for the bulbs in question. Bear in mind the proportions of the display—small flowers will look puny and insignificant by themselves in a large container, while taller species need a larger, heavier tub or they will become unstable when in bloom.

Part-fill the container with fresh, good-quality, peat-free potting mix. When it comes to soil depth, the size of the bulb and the resulting plant plays a

part, but bulbs generally tend to produce a lot of roots, so aim for at least 4in (10cm) of potting mix beneath them. This is more critical than the thickness of the layer of potting mix on top.

Pack in the bulbs. Normal spacing can be disregarded—just arrange them so they don't touch. The tops of the bulbs should sit at about 1¼–2in (3–5cm) below the rim, depending on their size. Fill to the top with potting mix and water well.

As the bulbs grow, water as necessary, but don't let the pots get saturated or frozen (see p.90). After flowering, feed regularly with a high-potash organic liquid fertilizer, such as tomato food, to help build up the bulbs for blooms next season. Bulbs in containers will diminish with time, and the smaller they are to start with, the more quickly they will fade, so ideally, plant them into open ground when they go over.

Specialist subjects

A lot of garden bulbs originate in the Mediterranean or Middle East, and although some are hardy, others may need shelter, excellent drainage, or both. Many grow wild in mountainous areas, benefiting from an insulating layer of snow in winter and flowering when the spring thaw comes. As garden plants they may thrive, but in a temperate climate there may be persistent rain, or frost without snow. The more demanding bulbs, therefore, lend themselves to container growing, as the plants can be moved into the protection of a greenhouse or cold frame if the temperature dips, or in periods of very wet weather, which provides more control over the conditions.

Top right The cool blues of *Anemone blanda* and dwarf irises combine nicely for a late-winter display.

Top far right Recycled enamel containers can be filled with bright seasonal bedding, such as primulas, and dwarf narcissus 'Tête-à-tête.'

Bottom right Snowdrops look best in a planter that is in proportion to their size, and shown off from an elevated position such as a garden table or the top of a wall.

Using Hard Landscaping

The purposeful design and creation of gardens is what sets them apart from natural landscapes, and while plants are a fundamental part of the appeal, the hard landscaping plays an important role in how the outside space is used and how well it performs.

Hard landscaping is an umbrella term that distinguishes all the materials and structures in a garden from the planting. On one level, it is functional; it defines the edges and makes a space usable. But it can also be something that is integral to the design, enhancing both the feel and the experience.

In winter, this is more important still. Even the largest, woodiest, and most evergreen of plants will fluctuate over time, but hard landscaping does not. It is the solid part of the garden, made up of items that are manufactured or built. It provides a frame for the plants and a gallery in which to display them, and when the hard elements are woven together skilfully, they can provide context, informing the way the space is used and perceived.

Once the greenery is gone and the perfidious perennials have retreated underground, the hard landscaping will still be there in full view, so it is well worth giving it the attention that it is due.

The Advantages of Hard Landscaping

Gardens come in all sizes, styles, and configurations, but judicious use of hard surfaces and structures alongside the plants can create attractive contrasts and make outdoor spaces both more usable and more enjoyable, too.

On an aesthetic level, winter is when the truth is revealed and good proportions and coherent design come into their own. Like the intricate mechanism beneath the bodywork of a machine or the elegant framework of a cathedral roof, hard landscaping can be a revolutionary force, supporting structure and beautiful in its own right.

No matter how discreet the hard surfaces are in summer, in winter they are conspicuous. Rather than being something ugly that must be tolerated until spring, therefore, hard landscaping should look good and work well alongside the planting, creating texture and enhancing focal points, in addition to making the space function practically on an everyday basis.

The different elements contribute in various ways. A path might add slick geometry or it could curve intriguingly toward destinations unknown; in either case, venturing outside is far more appealing with the prospect of dry feet.

Left Even in the depths of winter, well-chosen hard landscaping does not have to be dominant or in any way ugly, and paving, sculpture, and seating can work seamlessly as part of the overall design.

Boundaries don't just define edges, they contribute to the atmosphere and add privacy and shelter, while raised beds create growing space and allow plants to be appreciated at a convenient level.

In an ideal world, hard landscaping marries beauty with functionality. The path to the shed can be decorative as well as practical and steps make a wet or icy slope navigable. Stepping stones across a lawn will do away with muddy tracks and compacted soil, and they can also be used in borders for ease of access and care.

Permanent features can crystallize a design, becoming more obvious and revealing their potential as the leaves fall. A raised pond will provide a focal point in a formal garden and sculpture of any kind will revel in the extra quiet space that winter brings.

Even a garden room, greenhouse, or shed can be generous in its possibilities. Whether well-used or not, it represents somewhere to retreat, and a place to work, rest, or hide from the world; it is a place to potter about, sow seeds, and store things, as well as being a striking and substantial structure within the garden itself.

Challenges and Pitfalls

Getting the balance right between the planting and the built environment is key. Some backdrops are naturally lovely, but not everyone is blessed with mellow brickwork and weathered timber and it is common to have an outlook that is less than attractive. With time, however, hard landscaping can often be upgraded, while good planning at the outset enables many issues to be avoided when building anew.

The places in which we live and garden are full of hard surfaces, and these can be both a blessing and a curse. Paths, steps, and outbuildings can make a garden more usable in winter, but these should be chosen carefully, not just to avoid building in new problems but to ensure that the aesthetic is one to make the heart sing, rather than an expensive mistake to be repented at leisure.

In addition to this, few homeowners have a seamless and glorious surrounding landscape and many are challenged by views over which there is no control. These might include plain, tired walls, the insipid brickwork of a new development, or a neighbor's decrepit garage. The sparseness of winter and flat, gray days can compound the issue, but there are a range of solutions that can help improve the situation, and a long-term perspective on planting is useful.

Potential issues

Unattractive or dominant features: in young or sparsely planted gardens, the hard surfaces often take up more than their fair share of space, and mismatches will be obvious. A limited palette of materials can help, both when installing new features and when adding to what is there. Existing schemes can be refreshed or unified with a lick of paint or some form of cladding, while using additional plants to hide ugly surfaces can make all the difference.

Slippery surfaces: mud and algae can make hard surfaces slick and dangerous. Reduce the chance of accidents in well-used areas by choosing textured paving or adding grip-strips to decking and steps.

Poor drainage: rainwater will run straight off concreted or paved areas, while earth compacted by footfall or for artificial turf is also fairly impermeable. Pooling water and local flooding can be ameliorated by installing soakaways and additional planting areas, and also by improving the soil. Ponds and bog gardens can also be used to collect excess water.

Environmental impact: human-made materials always come at an environmental cost, so repurpose existing materials where possible. Stone must be quarried and is often transported thousands of miles, concrete releases carbon, while artificial turf smothers wildlife and makes unnecessary use of plastics that, at time of writing, are not recyclable. Timber can be the greenest option, but check how sustainable it is, or buy recycled.

Cost: compared to planting, landscaping materials are expensive and installing them also comes at a price. It may make economic sense to team costly stone with areas of cheaper gravel, or to look at reusing or upcycling materials that are already in your garden.

Right Hard landscaping dominates this newly constructed garden, but in a few years' time, when the plants have matured and filled out, the space will feel far more cohesive.

Choosing and Using Hard Landscaping

Hard landscaping is the Goldilocks branch of garden design. Too much will dominate aesthetically, cost a lot, and have a bigger impact on the environment; too little risks skimping on the essentials of paths and seating and will make the space feel mean and small. But if you get things just right, the garden will be both practical and attractive.

With the plants dormant and the extent of the boundaries visible, winter is often the time of year when gardens are redesigned or revamped. When planning a project, there are a number of things to bear in mind:

Be generous with proportions. Paths need to be wide enough to walk down comfortably and safely; patios and areas of decking must be sufficiently large to accommodate family and guests with ease.

Restrict materials to no more than three or four materials, finishes, or colors. Using a limited palette will help to achieve a cohesive look, but avoid painting everything the same color, as that can look clinical or bland. Contrasts are good for adding depth.

Tie the space together with a consistent theme, or blend new materials in with existing finishes. If there are wooden raised beds, use wood for decking too; if corten steel is a leitmotiv, stick with that, as adding other types of metal structure will look uncoordinated and haphazard when all are seen at the same time.

Seek green solutions when planning hard landscaping. Factor in transport, carbon, and water management, as well as considering products in the context of their whole lifespan, including the ease of recycling. Reused, recycled, and local materials are more sustainable.

Drainage is key. If a garden is entirely paved over, rainwater has nowhere to go, resulting in overloaded public drains and, ultimately, flooding rivers. Wherever possible, incorporate permeable surfaces into your design, such as loose-laid paving or stabilized gravel, and install rain barrels to collect water from shed and outbuilding roofs. A large, densely planted border will also help mitigate the risk of flooding by soaking up the rain.

Use planting to fix inherited problems, such as sheds, patios, or ponds that may not be a priority for upgrading or removal on grounds of cost. Winter is the perfect opportunity to develop new, cost-effective planting ideas to soften useful but ugly structures and improve the aesthetics.

Paths, patios, and paving

A good garden path is a thing of beauty, and whether it is crisp, straight, and purposeful or discursive and gently meandering, it makes sense to capitalize on its assets and potential.

A rustic bark-chip path edged with logs is ideal in a woodland garden, while informal stepping stones across a lawn break up the expanse of green and avoid a muddy track. Gravel is versatile and cost-effective and can be loose or resin-bonded. Brick and stone are fairly expensive options, but they are smart, durable, and ideal for areas that are heavily used.

Clockwise from top left A trained pyracantha softens the brickwork. A herringbone-patterned brick path adds a nice detail. Walls constructed with local materials contribute to a sense of place. This crinkle-crankle wall is just one brick thick, but its regular meander means that it is both stable and exciting to look at.

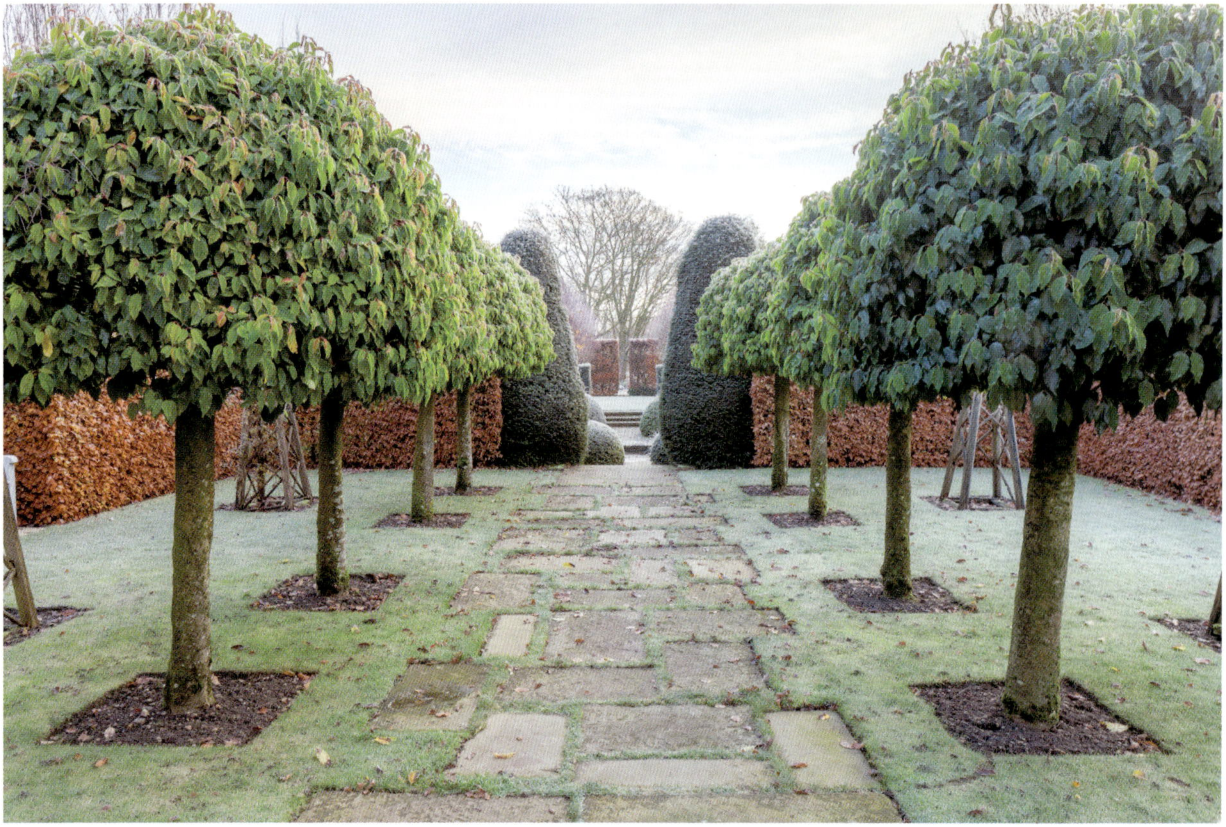

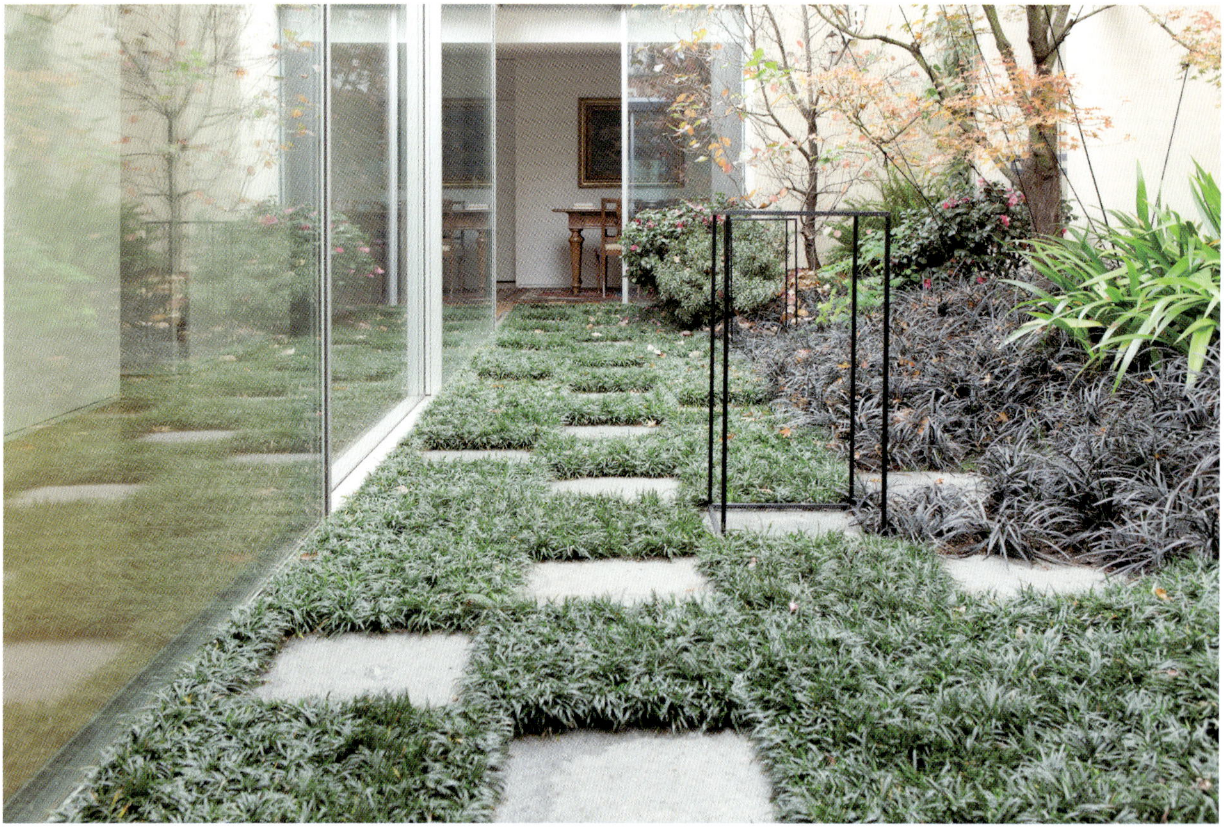

Success with paths and patios

- Use a material that is appropriate for the level of use and works harmoniously within the site.
- A neat, decisive edge defines boundaries and holds the design in place, both visually and practically, as well as helping to manage the transition from one area to another.
- Lay slabs on a gradient to stop water pooling, or allow it to soak away by leaving gaps or using a permeable substrate such as gravel.
- Think about what a path is for, its purpose and destination. A path can describe a journey, which is delightful if that is appropriate, but people in a hurry use the most direct route, and if this is not taken into consideration they will cut corners.

The width of a path can be determined by its purpose and footfall. For occasional use or aesthetic appeal, a path can be quite narrow, perhaps 20–24in (50–60cm) across—approximately the width of a person. A main route leading to a gate or doorway, meanwhile, should be at least 4ft (1.2m) wide—sufficient for one person to push a wheelbarrow or carry shopping bags in each hand, and for two people to walk side by side.

When it comes to wider expanses of hard landscaping, patios, decks, and driveways all need to be large enough to perform as intended, clearly, but they should also look good. Once again, the choice of material is important. Although a clean, pared-back effect is elegant, incorporating a pattern, change of pace, or decorative effect with bricks, cobbles, or mosaics will increase the visual impact, particularly in winter when the feature becomes a highlight.

Top left Weathered stone slabs work well in a traditional and formal design. Here, they add a low-key purpose to the scene.

Bottom left In a modern garden, checkerboard paving interplanted with low-growing evergreen *Ophipogon japonicus* 'Nanus' provides year-round interest.

Walls, screens, and boundaries

The vertical surfaces within and around a garden become far more prominent in winter. As the deciduous planting relaxes into hibernation and the last bright leaves drop from the climbers, the nearby buildings, fences, and boundaries assert themselves on the view. Whatever they are made of, and whatever color they are finished in, they are very much present and unavoidable.

Rather than being a disadvantage, however, this can be seen as an opportunity to look at the bones of the garden within its frame and context and to find ways to work with the underlying geometry. It is also a chance to look upward and outward at the surrounding houses and landscapes. Small things such as picking up the terracotta hue of a pantile roof, the plummy-gray of slate, or the inky-green of a distant pine wood can all settle the garden into its place.

While too many bare surfaces can be oppressive, adding vertical panels to gardens which are resolutely open can actively improve both the view and the feel of the space. A wooden fence or a brick wall contributes height and privacy, while screens may be incorporated on their own merits to create zones or to act as an intentional interruption to a linear feature, such as a hedge or border.

As a backdrop to structural planting, walls act as a foil and a canvas, as well as supporting trained shrubs and climbers to make best use of space in smaller plots. Yet they are an asset in other ways too, providing a microclimate that is both drier and warmer than elsewhere in the garden.

Nestled close to a building, plants that like a sheltered spot or drier conditions can sit out the winter quite happily, while the high thermal capacity of masonry means that it absorbs and retains heat during the day and releases it at night, increasing the average temperature nearby. This gives tender and marginal specimens some of the protection they need to overwinter outside, and enables other plants to flower sooner or to perform over a longer period.

Raised Beds

Simple and versatile, raised beds are an effective solution to a range of gardening challenges. In addition to creating space for planting, they can provide structure, definition, and seating; they are also easy to build so lend themselves to retrofitting into existing gardens as well as playing a part in new designs.

A well-designed system of raised beds can pull its weight in almost any garden situation. At one end of the scale, they can be a simple combination of boards or cubes, dividing up a space and defining a cultivated area, such as within a kitchen garden; at the other, they allow gardens to be created where there is no soil at all. On a steep site, meanwhile, they can be used to create terraces, helping to make sense of the changes of level.

Where local soil conditions are unfavorable, a raised bed is an easy way of cheating the site and growing plants that would otherwise struggle. In gardens on chalk and limestone the bed can be filled with ericaceous compost and used to grow acid-loving plants, while on heavy, wet soil, it will provide well-drained conditions that will be far more congenial for Mediterranean herbs such as lavender, and anything else that prefers dry feet.

Planted well, the raised bed literally adds another level to the winter garden experience. Rather than looking down on shrubs and grasses, it is possible to walk through an elevated landscape; small trees become magnified in bulk and height, as the new view from below the canopy alters the way in which they are perceived. In a season of low-growing loveliness, meanwhile, building a tall structure or creating a sunken path means that the small plants growing on top are closer to eye height and can be appreciated with ease.

Tips for success

Materials: these should suit the rest of the garden. Timber is cheap and versatile, and can be painted or left to weather naturally. Metal gives a slick, formal finish, while brick, stone, or even concrete can have vernacular charm.

Height: while some raised beds are just glorified edging, in winter the structure needs to be reasonably substantial. Both immersion in the planting and minute appreciation of small flowers require height and elevation, and bulk is necessary in order to deliver the benefits of increased thermal mass (see also p.93).

Drainage: given that a raised bed is essentially a large container, it needs a soil mix that will retain plenty of nutrients and water, while also draining well during long periods of rain. A mix of loam and organic matter is a good basis; if the substrate beneath is heavy soil or concrete, or it is a high rainfall area, consider adding some coarse sand.

Planting: think about how the bed will look in winter and actively design it for impact. Using plants with several seasons of interest will provide height, movement, and color all year round. Alternatively, use the well-drained conditions to create an unexpected display of xerophytes, or plants from the southern hemisphere countries such as Australia and South Africa, which are often well adapted to arid conditions and even drought.

Top right A raised bed incorporated into a wall creates warm well-drained conditions and allows less hardy plants to thrive.

Bottom right A new scheme can be designed with bespoke modern planters.

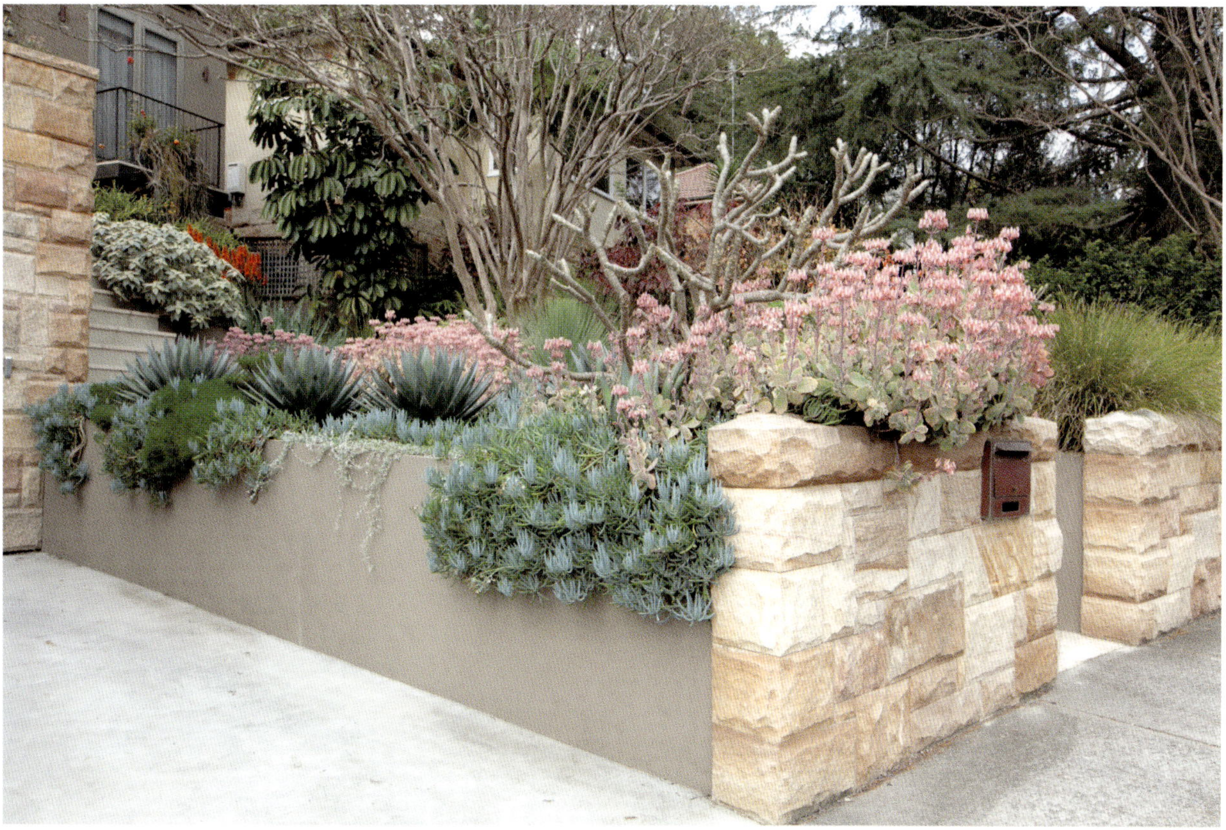

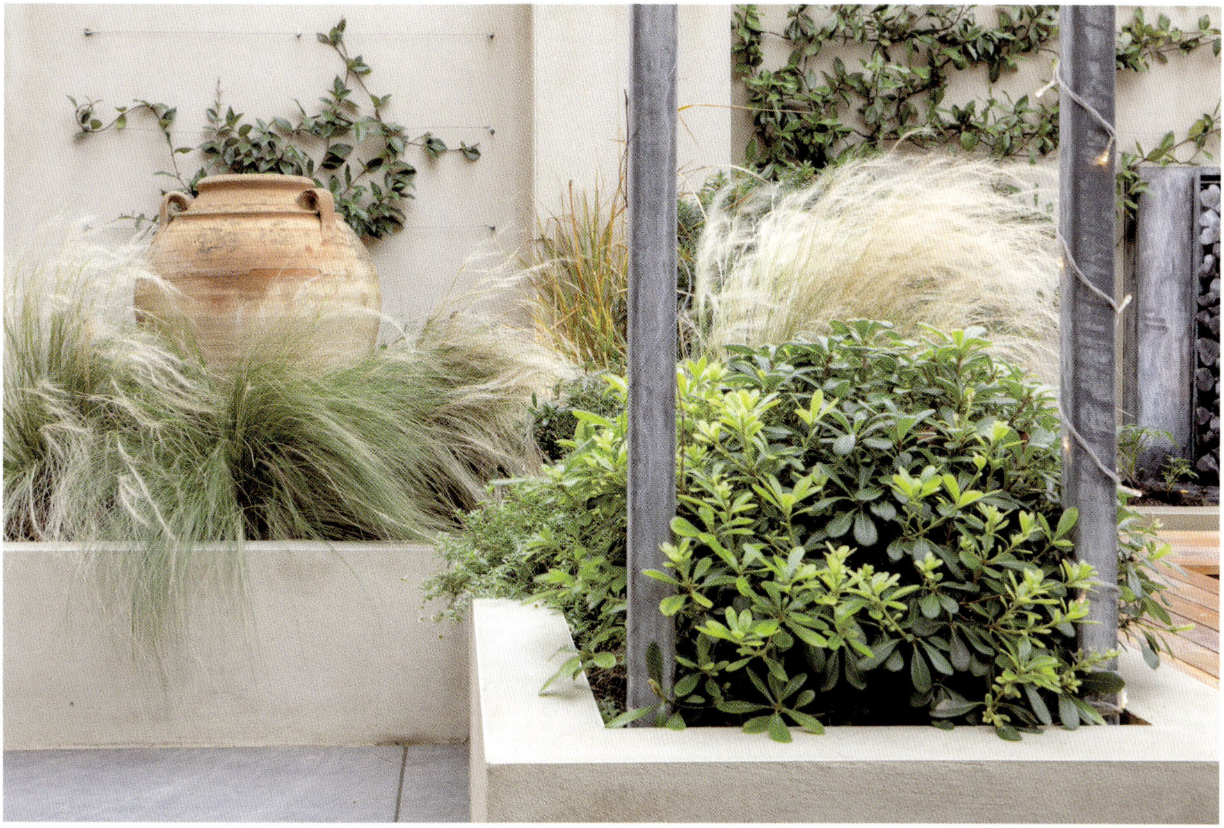

Garden Buildings and Structures

Pleasant retreats and useful necessities may not appear particularly prominent when they are first conceived, but as foliage ebbs, so pergolas, arches, and plant supports are thrown into the limelight. Greenhouses and garden rooms are writ unexpectedly large upon the landscape and even a shed that is tucked discreetly at the end of the garden for most of the year may suddenly become the focus of attention.

In the winter garden, the most prosaic structures can be thrust from supporting cast to star turn. There is no less interaction with the plants that grow around or up them, but the balance of emphasis switches and the potential for framing and geometry brings something additional to the design.

It pays, therefore, to plan for coherence and good looks at the outset. When buying a shed, make it a pretty shed. When siting the greenhouse, think about how visible it will be at all times of year, and how it sits in its landscape, as well as the usual considerations, such as putting it in a bright spot.

A gazebo may be a shady seat with a view, but when the scenery is thinner it might also catch a low winter sunbeam or at least provide shelter from the rain, while the garden room bought for working from home can offer a new, inviting, and warm perspective on winter-garden living.

Right It is easy for utility to outrank aesthetics when it comes to garden buildings, but when summer is past and the balance of interest shifts, the structures that are then revealed should be beautiful in their own right.

Features, Furniture, and Focal Points

In the process of creating a garden, it is usual first to plan the layout, landscaping, and planting. Once that is in place, however, the next pleasurable task is to think about the details and highlights within the space, and the many ways in which it can express the tastes and interests of the people to whom it belongs.

The way a garden is used and enjoyed is fundamental to its design, but within every grand statement such as this, there are subtleties. There are the practical details of timing: a vista might be planned with a focal point at the end, but as the scheme makes it off paper and into reality, it may be only later that the precise choice of feature is made. When it comes to creating a link with the house, meanwhile, assessing the way in which this is managed, and deciding how best to capitalize on the connection, can take thought and deliberation.

There may be visions to fulfil and big dreams to realize. Scouring salvage yards, showrooms, and the internet is all part of the exciting process, as is hunting for the perfect accoutrement, something that is just right and lives vividly in the mind's eye. A handsome watering can—or three—or the right birdfeeder may be the icing on the cake, and if your tastes run to green men, Victoriana, or steampunk metalwork, this is very much the point to indulge those whims.

Whatever the situation, the whole delightful process of "finishing" the garden and furnishing it in a way that is both intriguing and aesthetically appealing can make a substantial difference to its overall effect.

Objects of beauty

Offsetting the planting and adding personality, sculpture and other intriguing prospects are not just an expression of taste, but also distract the eye and create a more rounded and enjoyable experience. The purposeful addition of an obelisk, a piece of modern art, or animals cast in metal or stone make the space vibrant and energetic, while elements such as firepits really come into their own on cool dark nights, injecting warmth, excitement, and a party atmosphere.

A water feature is always an asset, and whether the budget stretches to an antique stone fountain, a raised pool, or a container pond, the effect is compelling. The sight and sound of trickling water is soothing and fascinating, as is the wildlife it will attract, and while the structure itself may be inherently beautiful, the freezing and thawing of the water, and the splashing of raindrops into its surface, has an appealing dynamic quality to it, too.

Finally, add plenty of seating. Where there is a secluded corner, there should always be a convenient chair in which to enjoy a moment of quiet. The interactive nature of a designed space should also be acknowledged, and where there is a view, or a highlight upon which to rest the eye and meditate, this necessarily requires a bench of contemplation.

Clockwise from top left Striking obelisks provide punctuation and height. A simple water feature creates an effective moment of stillness and calm. Plants can be pruned and trained to create the most glorious winter features. Sculpture, water, and clipped evergreens is a classic combination. Topiary brings theater to the garden. Chosen well, even a simple seat can become a garden feature.

Planting *for* Structure and Texture

Winter is a time of reckoning in the garden. When all the frills and foliage are gone and the herbaceous planting has faded away, its real character emerges, and what is left behind must carry the space until spring. The key to a strong design is, therefore, plants that perform when they are in full fruit or blossom, and also when they are not.

The form and character revealed when the leaves fall is important, and there is skill involved in building up layers of interest, and choosing plants for their architectural qualities, subtle details, and powers of reinvention.

Good form takes a garden to a new level. Trees are used for height, but their spreading quality is very different to the intense, whooshing verticals of bamboo. Color and tone, meanwhile, assume a new significance, with monochrome contrast or blocks of flaming color. Patterns and textures are deeply satisfying, while shapes can become a meditation on repetition.

Anchoring in summer and fundamental in winter, structure should be the first consideration when planting anew, and it is never too late to look at an existing garden with fresh eyes and thoughts of improvement.

Structure and Height

In the stark, exposing clarity of winter, a garden relies on its structure. It is the core element, bringing purpose, strength, and crispness, while underpinning the architecture of the space during the rest of the year as well.

When the bones of the garden are revealed, it comes into its own. With the edges of the space and the symphony of the planting within it all visible at once, it is finally possible to get a real feel for how it works.

Herbaceous perennials will inevitably collapse, but if they are backed up by good structure, it is barely noticeable that they have gone. In their place, there are plants whose organic geometry provides framing and bulk; plants that are angular but beautiful or that add bounce and rhythm, height, and richness, regardless of the season.

As key structural plants, trees and shrubs are an important part of a garden at all times of year—but our relationship with them is complicated. People worry about their potential size and fashion has played its part in fostering ambivalence or outright avoidance; conversely, bedding plants offer temptation without commitment, while small gardens often dupe their owners into downsizing their contents unnecessarily.

Left Vertical accents and repeated shapes add energy and excitement to the planting scheme, while judicious use of evergreens and contrasting layers ensures that the garden retains a sense of fullness, even in frost.

But it is worth remembering that the distinctive landscapes of older gardens are largely based on mature plants, and the most enviable and exciting of those are usually woody and structural. Perennials are pleasant and bedding has its place, but they are not imposing, and they will never provide the size, longevity, and year-round interest of a well-chosen tree or shrub.

Indeed, tall plants are a real asset to a design. They draw the eye upward and outward in an expansive manner, while the bulk of bushes and small trees can also be used to make a space feel larger. Combining and layering structural specimens plays with perspective to create a sense of depth and mystery. Corners can be hidden and discovered, bringing in another dimension that is particularly useful in a small garden.

And while the ultimate size—both spread and height—should always be a consideration, in reality there are relatively few woody plants that can't be pruned or trained to improve their appearance or to keep them compact and under control.

Introducing Drama

When it comes to architectural plants, winter does things differently. Rather than a parade of big-leaved beauties, it capitalizes on an innate elegance of form: the trees that are sculptural in their nudity, the stems that make a statement, and evergreens in all their manifold glory.

In pursuit of the bold, the unexpected, and the theatrical, architectural plants can be used to create a scene in which size and scale are almost irrelevant. Magnificent cedars or substantial topiary shapes are lovely indeed, but small spaces can also enjoy a sense of drama, with containers of clipped lollipops on each side of a door, or a cluster of tall pots filled with succulents, phormiums, or evergreen grasses.

A jolt of surprise and a sudden prickle of interest can be achieved through a sharp contrast or change of pace, but the theatricality can also be sweetly harmonious. Rather than a crisp frame, a focal point might emerge from a sea of softer companions to poetic effect, while moderating the idea of drama and stitching it into an existing theme means it can be tailored to suit any style and taste.

Drama is more than assembling the building blocks of structure. It embraces architectural plants with the panache to put on a real performance: a handsomely proportioned specimen tree with its assembled cast of supporting elements that conspire to make it the focus of attention. It is about creating a narrative and telling a tale through the medium of strong and thoughtful design.

Right When dressing the garden stage for winter impact, repetition is an important tool. In garden design, as in music, it creates a regularity and rhythm that has a comforting and mesmerizing effect.

Pattern and Texture

When the verdant cacophony of summer and fall is done, subtleties that were drowned out by planting or obscured by foliage come to the fore.

Ancient trees with reaching branches and craggy trunks are silhouetted on the skyline, and the garden jostles with quiet diversity. Erratic *Corylus avellana* 'Contorta' competes for attention with the enameled copper bark of *Prunus serrula*, while smooth, gray ash and dappled birch combine elegant texture with soft, watercolor hues.

Pattern

The winter garden is full of shapes and patterns, which contribute enormously to the visual effect, yet the lines between crafted and natural are often blurred. The shaping of plants usually draws heavily on their natural form and behavior, and where there is repetition within a design, the work of the human hand is overshadowed, to a great extent, by a tendency toward repetition, geometry, and the inherent artistry of the natural world.

Some shapes and patterns are large and distinctive. The overriding impression of bamboo is its verticality, although within this it can also be segmented or kinked. A tree, meanwhile, can be appreciated in all its squat, pendulous, spreading, or gangly glory.

Clockwise from top left The honey-hued bark of *Betula albosinensis* 'Bowling Green.' Marbled *Arum marmoratum*; *Stipa tenuissima* contrasts well with the thick leaves of *Phlomis*. The twisty twigs of *Corylus avellana* 'Contorta.' Mesmerizing *Euphorbia myrsinites*. Old man's beard, *Clematis vitalba*, scrambles through a wild winter hedgerow.

On a smaller scale, the Fibonacci sequence (p.56) is much in evidence in botanical spirals and repeating rosettes, from saxifrage to sempervivums, while the manner in which leaves organize themselves on a stem is a meditation on use of space that can be seen in every mahonia or evergreen fern, and is at its best with a dusting of frost.

Variegation, too, plays its part. *Ilex aquifolium* 'Argentea Marginata' and cultivars of *Pittosporum* have a rimed quality, while the mottled, blended leaves of *Cyclamen hederifolium* and *Arum marmoratum* echo the marbled glass of an Art Deco lampshade.

Light and cold weather have a huge impact on the garden in winter, and patterns are part of this. With low light levels and snow, the garden can look like an exercise in modern art: Cubist geometric shapes, evergreens that seem block-printed onto paper, and twigs scribbled across the sky like graffiti. Meanwhile, the softer lighting, with mist or frost, generates an impressionistic or conceptual feel.

Texture

For many plants, the winter reinvention is not just visual but tactile, too. *Acer griseum* is famous for having polished bark that flakes away in dramatic curls, snakebark maples such as *Acer davidii* and *A. capillipes* sport vertical striations against a colored backdrop, while the winged twigs of *Euonymus alatus* are both unusual and unexpected. *Stewartia pseudocamellia* and *Parrotia persica*, meanwhile, resemble marbled wallpaper, and the trunks of certain pines and redwoods develop a softly shaggy outer coat, rather like an arboreal yeti.

In a delightfully sensory world, the peeling cinnamon stems of hydrangeas demand investigation and clematis seedheads hover like puffs of silver, teasingly out of reach; conversely, the tempting thorns of frosted Japanese wineberry, *Rubus phoenicolasius*, and bewilderingly three-dimensional leaves of hedgehog holly, *Ilex aquifolium* 'Ferox Argentea,' illustrate the allure of things that should not be touched.

The History of Topiary

Popular in gardens for thousands of years, topiary is the art of clipping plants into a range of sculptural or decorative shapes. Providing year-round structure and interest, they are often a useful backdrop or focal point as well.

In Europe, topiary first appeared in Roman times, with Pliny the Elder describing evergreens clipped into the shape of obelisks and animals in the Tuscan villa belonging to high-ranking dignitary Gaius Matius. The practice spread beyond Italy and has achieved peaks of popularity at various points in history.

Following the Norman Conquest, the British became enthusiastic about mazes and labyrinths, while the later Tudors planted knot gardens, with "ribbons" of different-colored box in a crisscrossed or knotted style.

Parterres made of clipped box and yew, in repeated swirling patterns or orderly geometric designs, took off during the Renaissance and remained popular in Europe, particularly in the châteaux of France. In Britain, the practice of topiary was revived in the late 19th and early 20th century. This acted as a formal foil to the relaxed planting in Arts and Crafts gardens, as epitomized by the work of Gertrude Jekyll and Edwin Lutyens, and due to its enormous versatility, it has remained in fashion ever since.

In parallel to the European tradition, topiary has also been practiced in China and Japan, but rather than rigid geometric forms, the aim here is to create trees that represent an idealized view of nature.

Right Spectacular topiary towers make for a moody scene in the gardens of Château de la Ballue, Brittany, France.

Types of topiary

While topiary is a general term for clipping and design, within this there are several groups.

Architectural topiary: these are bold and simple forms that add structure and hold the garden together in winter. Including clipped evergreen hedges, together with cones, balls, blocks, buttresses, and other shapes, this is possibly the most adaptable form of topiary as it can be scaled up or down to suit gardens of all sizes.

Sculptural topiary: works of art that can be as formal or as fanciful as one could wish, these range from neat spirals to layered towers, birds, and other animals, and even theatrical scenes, all of which are lovingly pruned into shape over a period of several years.

Garden features: these are elements such as knot gardens, parterres, mazes, and labyrinths, which are usually intended to be a freestanding part of the garden—or even the theme of a whole garden. They are often designed to be appreciated from a certain angle, such as looking down from a vantage point, or walking between or through the topiary forms.

Cloud pruning and bonsai: possibly the best known of the varied and interesting Asian pruning traditions, Niwaki is a sophisticated Japanese cloud pruning technique. Creating a soft, rounded effect, it is more naturalistic than other forms of topiary, and lends itself to the stylized landscapes of Japanese or Zen gardens, along with Karikomi, which is used to create billowing, cloud-shaped hedges. Bonsai is similar to Niwaki, but the plants are grown in containers.

Trained climbers: a quick, satisfying, and fairly inexpensive way of getting the effect of a topiary form is to train evergreen climbers over a wire frame, and then keep it tightly clipped to shape. This can be adapted to suit both large and small spaces, and lends itself to inexperienced or time-poor gardeners.

Creating topiary

- Suitable plants include small-leaved evergreens such as yew, box, privet, or *Ilex crenata*, although others can be used, including rosemary, and also beech, as while it is technically deciduous, young twigs can keep their leaves in winter.
- Simple cones, balls, or blocks are the easiest shapes to start with. A standard or lollipop form requires the plant first to be trained to provide a single upright stem, and this could take several years. Larger plants can be cloud-pruned retrospectively.
- Use a template to help create a regular shape such as a cone or ball, and start carefully clipping. A rope wound around the plant can be used as a guide to make a spiral, while cloud-pruned forms can be done by eye.
- As the plant grows, the regular maintenance pruning will cause branching, making the shape more dense. Don't cut off the top shoot, or leader, until the piece is as tall as you want it to be.
- Once the topiary shape is finished, maintain it by clipping annually, or biannually if it is fast-growing. The precise timing depends on the plant in question, with box usually clipped in early summer, and with yew in late summer.

Top right Cloud pruning creates dramatic effects. While evergreen plants are common, beech can also be used as the leaves are retained in winter when clipped.

Top far right A beautifully designed, well-maintained parterre is mesmerizing both in its simplicity and in its detail.

Bottom right With time, topiary forms and clipped hedges can become magnificent and monumental features that add real character to a larger garden.

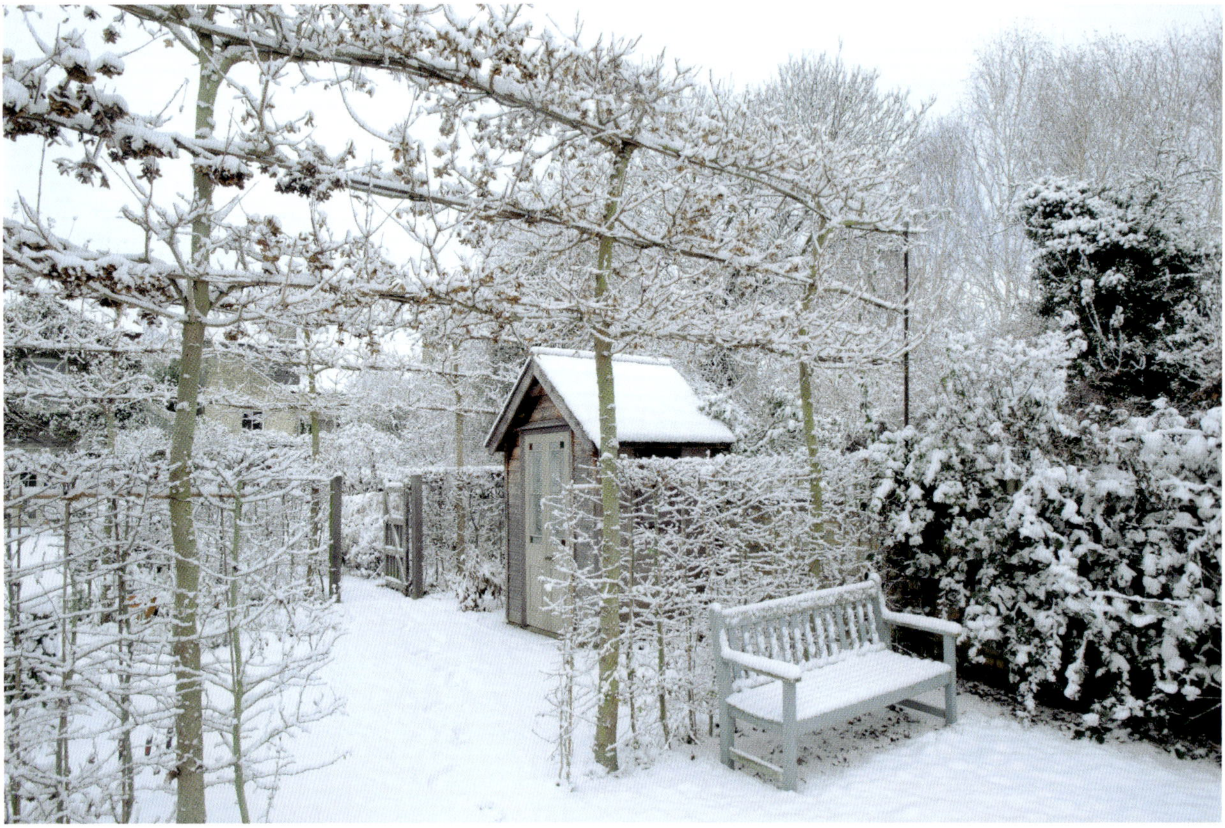

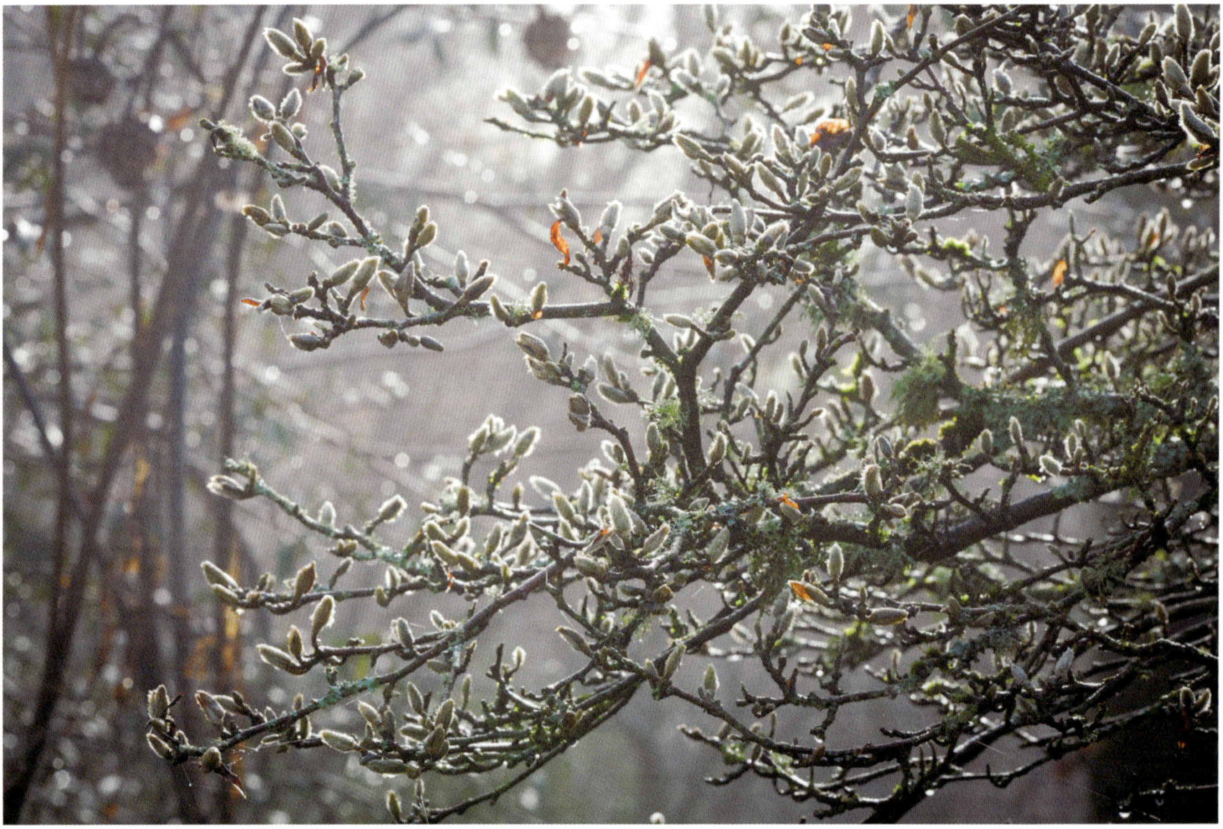

A New Season of Interest

Some plants are irrevocably associated with a certain month, but the brighter they shine, and the more surely they define their moment, the more likely they are to be ignored, often unfairly, at other times of year.

Many of the plants that are written off as one-season wonders deserve to be reviewed and rebranded as they may, in fact, have much more to offer. Saucer magnolias epitomize this: their spring flowers are glorious but ephemeral, so they have gained a reputation for being a plant for large, full gardens with plenty of other things going on. Yet, even when the flowers have gone, the spreading tree is a pleasant addition. It has a touch of fall color and low, snaking branches, while the new buds grow fat and fuzzy as the winter progresses. Seen as such, it does not so much define one season as connect them all.

For the plants that have a truly fanatical following, however, progress marches in their favor. Some, such as lilacs, have historically been bred for hardiness or large and varied flowers, but in shrubs, the trend is increasingly for compact size and an extended season of interest. New introductions and rediscovered gems may now flaunt leaf detail, stem color, fall interest, and an ability to rebloom that bucks the more well-known family trend.

Even within those groups of plants that have not yet been touched by the magic paintbrush of horticultural development, there are plants that deserve reconsidering. Rhododendrons are eye-popping in spring, and when the flowers fall, attention can turn to other qualities, such as the tactile, felted foliage of varieties including *R. pachysanthum* and *R. makinoi*, and the sinuous and mottled stems of mature plants, some of which can become positively arboreal.

With their notably long-running performance and multiple seasons of interest, trees contribute significantly to a strong garden. *Koelreuteria paniculata* has pretty yellow flowers, colorful foliage, and dramatic inflated seed pods, while mulberry and hawthorn develop craggy good looks and characterful form with time. Large shrubs such as *Amelanchier* with its delicately pointed buds, hydrangeas with their shock of naturally dried flowers, and even the bizarre and exotic seed cases of tree peonies contribute both form and additional interest, as the season progresses.

Climbers, too, may have a second coming. Though the tassels of lavender flowers are long gone, a well-grown wisteria will remain pleasing; clematis lives on through its seedheads, and rosehips can last for months.

Fruit trees, however, must be among the best candidates for a summer-to-winter translation. They may be grudgingly attributed two seasons of interest, in fruit and in flower, but rarely are they appreciated for their sheer garden versatility. Form varies enormously, from bold twiggy pillars and cordons to charmingly natural, spreading specimens. Trained as an espalier or fan they will cheer up a winter wall, and when mounted on wires, they make a great translucent screen or permanent backdrop to an herbaceous border.

Top left Trained fruit trees on tall trunks provide an elevated hedge and useful summer screen. In winter, meanwhile, they add valuable height and interesting structure without being overbearing.

Bottom left Backlit by the low winter sun, the silvery fuzz of saucer magnolia buds creates an effect that is infinitely more delicate than the flowers that follow.

Planting Shrubs and Trees

For a garden that is handsome and healthy, it pays to give new plants a good start so they can grow away strongly. Regular attention to their shape and form in the first few years also reaps dividends.

Woody plants are hugely varied, so before buying and planting a tree, review the conditions of the proposed site, as light, soil, drainage, and pH will all affect success enormously, as will the level of exposure and direction of the prevailing wind. The specifics of the plant are important too. If the plant has been grafted, ascertain whether the graft union (usually visible as a bulge or scar on the trunk) needs to be above the ground, as in fruit trees, or below it, as in lilacs and roses.

Plant both bare-root and containerized specimens to the depth they were growing before. Dig a hole that is big enough to accommodate the roots so that they can spread out properly, as this will anchor the tree. Avoid pot-bound specimens. Where a stake is necessary (for example, on a windy site), hammer it in before planting to avoid root damage.

Put the tree or shrub in the hole, teasing out the roots. If you are adding mycorrhizal fungi to support the tree's growth, do so now; then backfill with soil, watering well and firming around the base of the tree with your boot.

Finally, mulch around the plant with organic matter such as leaf mold or bark chips, making sure it does not touch the stem. This will help to keep water in the soil, and to reduce competition from other plants while it establishes. Water and feed as necessary, particularly in the first year.

Pruning

Most shrubs and trees will benefit from some pruning to keep them healthy and looking good, which is an important part of the winter aesthetic. It will also help to maximize the number of flowers or fruit. Young plants benefit from formative pruning, to help develop a good and balanced shape; once this is established, a maintenance pruning regime will keep them strong and productive. Below is an outline of techniques and timing.

Winter pruning opens up the shrub or tree and refreshes the wood for flowers in the future. At this time of year, the plant is dormant, so the wound can heal before the sap starts to rise again. This is particularly important in species such as maples and birches, which should be treated carefully and pruned as they head into dormancy, rather than as they start to awaken. Using sharp pruners, loppers, and a pruning saw, start by cutting out wood that is dead or diseased, then remove inward-facing branches to create a nice shape. Overlong branches can be reduced by a third to a half, if necessary. Shrubs with multiple stems arising from the base can benefit from having some of the oldest stems removed each year.

Pruning flowering shrubs is timed according to how long it takes for new wood and flower buds to develop, and the period of bloom. If flowers are produced during late summer, fall, or spring, cut back the plant in late spring or early summer so that it has enough time to produce new flowering buds. Avoid pruning during hard frosts.

Top right Established trees and shrubs can be wonderfully bold and sculptural.

Bottom right Leaving the spent flowers on mophead hydrangeas will protract their winter interest and protect the new buds. They can be deadheaded in early spring.

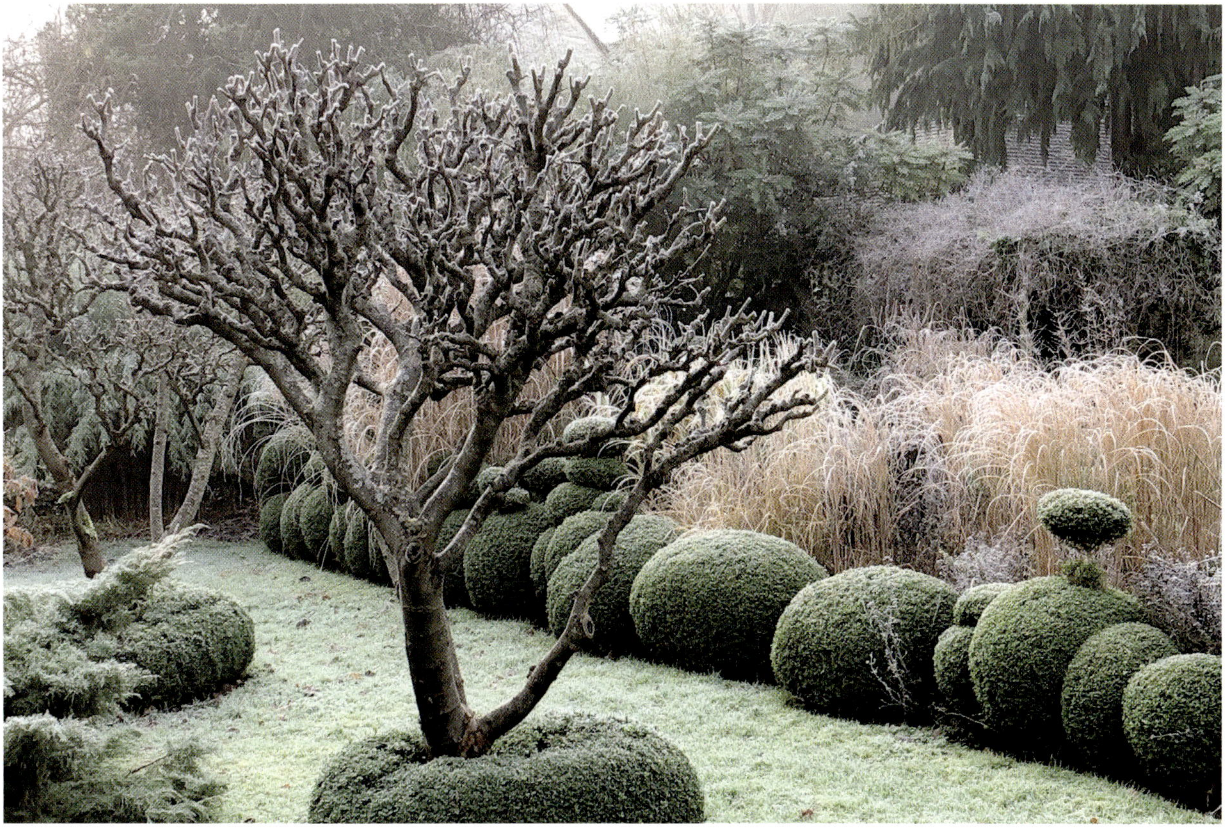

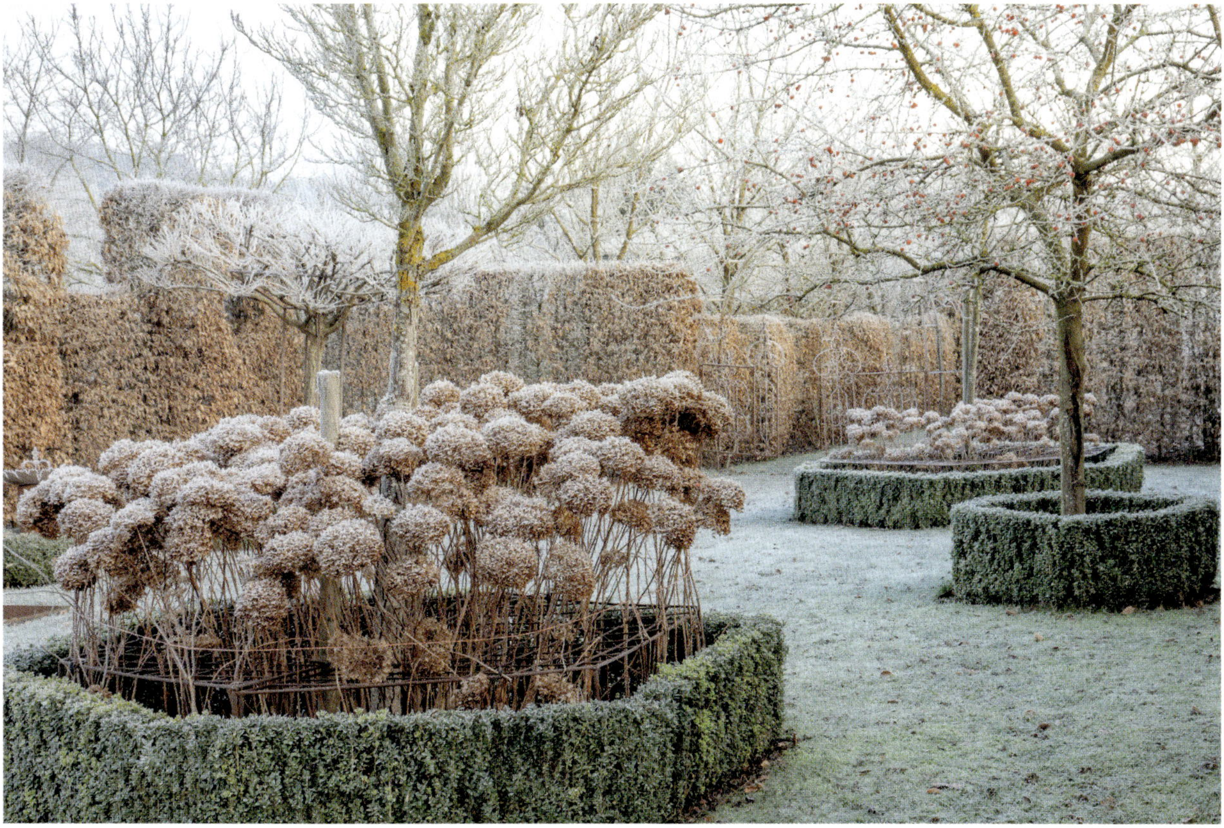

Colored stems should be pruned when buds burst at the end of winter. Cut out one stem in three at the base for a continuous supply of new wood; some established plants can be stool-pruned (coppiced entirely) every year or so.

Evergreens typically require little or no pruning, as they are often grown for their attractive natural shape. Where needed to control size, improve form, or maintain a clipped effect (see pp.124–27), prune after flowering or, if the flowers are not relevant, prune soon after the plant has started into growth to avoid bald patches.

Avoiding disease

In some cases, plant health and risk of infection is a factor that needs to be taken into consideration when timing pruning activities. This is why it is recommended that stone fruit such as plums and cherries are pruned in summer, to avoid infection with silverleaf disease, while cutting box in early summer is one precaution gardeners can take against box blight.

It is, however, important to remember that plants don't "need" pruning. It is done for aesthetic reasons or for the gardener to get a particular result, and in many gardens the counterpoint between plants that are clipped and those that are shaggy is an attractive one. So when in doubt, don't worry too much: just do some research and give it a go. You can always cut back more next year if necessary and, with a few exceptions, the worst thing that will happen if you prune at the "wrong" time of year is the loss of a season of flowers while the plant gets back up to speed.

Right A thoughtful approach to structure, texture, and repetition helps to create a garden that is satisfying to look at and sits well in its landscape.

Winter Bulbs

As a blaze of fall glory heralds the waning of the year, and cooler, shorter evenings hint at the dark days ahead, the bulbs that appear on garden center shelves are a chance to make ourselves a promise. They are a small step on the way to the flowers that lie ahead, and that will soon lift the heart, inspire faith in nature and reassure us that all is well with the world.

But there is no need to wait, for with golden leaves come fall-flowering *Crocus speciosus*, cyclamen, and colchicums. Snowdrops are the most anticipated flower of the new year, but this peak is preceded by an excitable flurry of early *Galanthus* species, paving the way for seasonal staples; not just the snowdrops, but perky crocuses and the winter aconites that sparkle like gold, the dwarf irises and the early daffodils not far behind.

Bulbs reward their grower with success and pleasure. While modest and humble, they deserve celebration, arising each year as a statement of optimism and a declaration that, whatever the winter can throw at us, there will be flowers waiting on the other side.

Making *the* Most of Bulbs

Bulbs are often thought of as exclusively spring plants. Snowdrops open the show, then almost without blinking, it's straight on to large-flowered crocuses, narcissi, and tulips. Yet this underestimates the true extent of their garden potential and diversity, and the length of interest that bulbs can provide.

A number of fairly diverse genera are known as bulbs for easy reference purposes, and many of these come from central and southern Europe, the Middle East, and Central Asia—areas that typically have hot, dry summers and cooler, moister winters. In fall, when the temperatures drop and rain becomes more frequent, a "second spring" arrives as plants put on a burst of new growth and bulbs come into bloom.

These fall-flowering species are closely related to familiar spring bulbs, within genera such as *Crocus*, *Galanthus*, and *Narcissus*, as well as *Cyclamen cilicium*, *C. hederifolium*, and *Sternbergia lutea*. Certain South African bulbs also flower in fall and into early winter, including *Hesperantha coccinea*, *Nerine bowdenii*, and various forms of *Gladiolus*. Therefore, with thoughtful planting and by employing a range of species and cultivars, there can be bulbs in flower right through from fall to spring.

In the garden, it is *Galanthus* that first commands attention in the depths of winter, becoming more numerous and varied as the season wears on, but

cyclamen are similarly robust performers, with hardy *C. hederifolium* giving way to *C. coum*, together with *Eranthis hyemalis*, and the early crocuses that arrive soon afterward.

What's more, with a little planning, it is possible to bridge the gaps and cheat the season. When kept in a sheltered location or under glass, pots of bulbs will often flower sooner than the same variety planted in the ground, while some varieties can be forced for an earlier display.

Fall-flowering bulbs

Species of bulb that flower in fall are often sparked into growth by the first rains of the season, so they can be encouraged to appear early by watering in late summer. In many cases, the flowers precede the foliage, either partially or entirely. Good subjects include cultivars of *Galanthus reginae-olgae* and the less widely available *G. peshmenii*, together with *Zephyranthes candida*, *Sternbergia lutea*, and *Crocus speciosus*.

Fall-flowering bulbs tend to prefer a sunny, well-drained site; feeding will help the plants to establish. Leave the foliage until it dies back naturally.

Left Flowering bulbs are an essential part of any winter garden.

Understanding Bulbs

Bulbs are plants in which leaves and flowers arise from an underground storage organ. When conditions become favorable, this provides energy for the plant to grow quickly, following a period of dormancy. It is, however, a term that encompasses not just true bulbs, but also corms and tubers, all of which are botanically distinct.

True bulbs are composed of layers of scales— thickened leaves that grow on a short stem called the basal plate, with the roots produced below. At the point when the bulb starts to grow, the new flower has already formed in the center of the bulb and will be at the top of the shoot as it emerges. Examples include snowdrops, lilies, and daffodils.

Corms are enlarged or swollen stem-bases sitting on a basal plate. Found in plants such as crocuses and *Iris reticulata*, these are covered in a fibrous outer layer called a tunic. Corms may have several growing points, each of which will produce a stem, and flower buds form on top of this.

Tubers are thick underground stems. They have no basal plate and bear buds, often in clusters at the top of the tuber, from which the new growth arises. Examples include *Cyclamen*, *Anemone*, and *Ranunculus asiaticus*.

Cultivating a passion for bulbs

Thanks to centuries of assiduous collection, the plants in our gardens come from all over the world and bulbs, in particular, have the power to fascinate. Objects of obsession and even mania, they have inspired plant hunters and enthusiasts to scale mountains, traverse gorges, and search scrubland in pursuit of the new, the rare, and the beautiful, with representatives of the bulb trade often hot on their heels.

Many bulbs are now commercially grown, but their far-flung ancestors evolved in a unique combination of climate, altitude, rainfall, drainage, and soil pH, and within a distinct community of surrounding plants. It is, therefore, unsurprising that they look different and behave in different ways, even if they are quite closely related.

This level of variety means that there are options to suit a wide range of gardening conditions and soils, but it also presents a number of variables. Success lies not just in understanding what each type of bulb will want and how it will behave, but what the garden has to offer—and finding the sweet spot between light, temperature, drainage, and water.

As a rule of thumb, bulbs like plenty of water while in growth, but when dormant they prefer drier soil as too much moisture can cause rotting. Good drainage is usually essential, and in some species, protecting the bulb from summer heat or winter cold may be necessary—replicating the effects of cooling altitude or an insulating blanket of snow, for example.

All gardens are different, however, and all plants need to have a degree of tolerance or they would not last long either in the wild or in cultivation, so it is worth experimenting, within reason. Wet feet can be offset by full sun; heavy soil rendered habitable by larger plants drawing out excess water; and sharp drainage ameliorated by summer shade (see p140). Any growing instructions are inevitably approximate and if you give it a go, the plants may thrive against the odds.

Right A selection of bulbs, corms, and tubers waiting to be planted.

Choosing and Planting Bulbs

When massed together, bulbs are a major event in any winter garden so it pays to make the most of them. Pick the best available varieties and plant as many as possible—in most cases, they will then come up cheerfully, year after year.

When choosing bulbs, there are a number of things to take into account:

- **Note the period of interest of the bulb.** In cool conditions with plenty to drink, a flower can last for several weeks, but displays can be extended by planting different bulbs to emerge in sequence. Some varieties such as *Galanthus woronowii* and *Cyclamen hederifolium* have leaf interest, too.
- **Plant attractive companions.** These will act as a foil to the bulbs when they are in bloom, and distract from leggy or tatty leaves after they have finished.
- **Work with your soil.** Clay soil can be very wet in winter but baked dry in summer, so consider moderating this with the shade and water absorption capacity of a deciduous tree. Sandy soils, meanwhile, struggle to hold moisture and nutrients, so improve them with organic matter.
- **Do not disturb.** Bulbs generally dislike disturbance, so if they are planted in the herbaceous border, the spot should be marked so when the perennials are divided the bulbs are not dug up too. Avoid grass paths and areas of high footfall and don't cut back or mow the growing leaves.

How to plant bulbs

As a general rule, bulbs should be planted at a depth that is 2–3 times their height and spaced the same distance apart, so a bulb ¾in (2cm) in height needs a hole that is at least 1½–2½in (4–6cm) deep. But use your judgment; in dry soils or warm sites, some

Size matters

Alongside growing conditions, the size of a bulb and its resilience to desiccation can influence success enormously. Small bulbs have a much greater surface area to volume ratio than large ones, which means that they lose water faster. Left in a warm room, or on a garden-center shelf, snowdrops and cyclamen will dry out much faster than substantial tulips or daffodils, and they may simply die. As a result, many small bulbs will favor shade while they are dormant as it provides the utopia combination of cool and stable conditions and soil that is dry, but not too dry.

species of bulbs or corms may benefit from deeper planting in order to stay cool and moist, but others, such as cyclamen, naturally grow near the surface and should be planted in the appropriate conditions, not more than 1in (2.5cm) under the soil.

Plant the bulbs with the shoot, or point where the flower will emerge, facing upward. In some plants, such as anemones and fritillaries, this is not always very obvious but plant them anyway—if you get it wrong they will still grow. Replace the soil on top of the bulbs and water them in.

In open ground, bulbs don't necessarily need to be fed, but an organic mulch will improve soil structure and can be used as a vehicle to add nutrients if necessary.

When planting in containers, all rules of spacing can be thrown out of the window and the bulbs packed in for impact, then planted out later. They will benefit from regular feeding (see p.143).

Top right With its sky-blue petals and bright gold markings, *Iris reticulata* 'Cantab' is one of many beautiful early-flowering irises.

Top far right Cultivars of *Narcissus cyclamineus* such as 'February Silver' (pictured), 'February Gold,' and 'Kaydee' often flower relatively early in the year.

Bottom right *Cyclamen hederifolium* produces pink flowers in fall, while the marbled leaves provide eye-catching groundcover for months.

Tips for success

- Healthy, fresh bulbs produce the best flowers. Purchasing early in the season from a reputable supplier reduces the chance of the stock having rotted, dried out, or been stored next to a radiator for several months.
- You will get better results if you pay attention to the natural preferences of the species in question, particularly its needs for light, drainage, and moisture. When you get this right, not only will the bulb not struggle in its first year, but there is a greater chance that it will survive and proliferate for the future.
- Generally speaking, the earlier a bulb flowers, the earlier it should be planted. Snowdrops need to be in the ground by early fall, while fall-flowering crocuses should be planted in late summer to thrive. Late-spring-flowering tulips, meanwhile, can wait until late fall or beyond, as it will still be months before they bloom.
- Water containerized fall-flowering bulbs in late summer to get them going.
- Feed bulbs in containers and borders while they are in active growth, using a high-potash fertilizer. Don't feed small bulbs in the lawn as it encourages competition from lush, well-nourished grass.
- Make sure that bulbs have good light during winter and spring, while the leaves are growing and photosynthesizing—they'll struggle in deep shade or under dense evergreens.

Using bulbs in the garden

Winter-flowering bulbs are hugely versatile and there is something for every situation. Bold and striking when planted en masse, they are ideal for providing a splash of color when all around seems to have paused for breath and, when faced with a tricky gap or a sparse planting scheme, adding a good dollop of bulbs is a quick and easy solution.

Create a spread: when it comes to impact in the winter garden, multiplication is key, and tall or double-flowered varieties can be used for extra landscape impact. As clumps form, snowdrops, crocuses, and narcissi can be dug up and divided; *Eranthis* and cyclamen, meanwhile, will self-seed if they are happy, and the seedlings can then be spaced out as desired.

As part of a tableau: plant bulbs such as cyclamen, scilla, and winter aconites in combination with seasonal companions such as hellebores and evergreen ferns. Underplant them liberally as a foil to set off deciduous shrubs, such as hydrangeas, dogwoods, and tree peonies, or to enhance the look of specimen trees.

In the lawn: lawns are often better left untrodden in winter but far from being boring, they make a great backdrop to an ephemeral planting of crocuses. These will vanish below ground by the time the space is being used. Plants that are leafier or have a longer growing period, such as colchicums, cyclamen, and erythroniums, can create a colorful pool around multistem birches, *Parrotia persica*, or a lovely old apple tree.

Container heaven: clusters of pots are a good way to create instant impact, either as single-variety containers, for example, a display of miniature daffodils and *Iris reticulata*, or in combination with *Ophiopogon planiscapus* 'Nigrescens,' winter-flowering pansies, grasses, dwarf evergreens, or even striking succulents.

Create a theater: elevate your display to a bespoke centerpiece, with tiered shelves to show off the potted bulbs. Alternatively, try staggering the height of the containers, perhaps by arranging them on boxes or upturned pots of different sizes, and interspersing with trailing ivy.

Left Planted into individual terracotta pots, bulbs such as *Crocus* and *Muscari* create a pleasing splash of color.

Interesting and Lovely Bulbs to Try

For creating a cheap and cheerful splash of color, winter bulbs are a universal favorite, but when dominated by commercial favorites the display can sometimes look a little predictable. Dig a little deeper, however, and you will find range of specialist suppliers with all sorts of choice plants to explore.

With flowers that grow so willingly from humble beginnings, bulbs are addictive. The obsession starts innocently enough: perhaps the catalyst is a pack of daffodils picked up with the groceries, followed by a vivid pink cyclamen, or the vintage charms of a forced hyacinth. But with pleasure and performance almost guaranteed, temptation then sets in.

Late-night internet searches and Sunday morning catalog trawls reveal daffodils to be almost infinitely diverse and snowdrops to have cult status. Meanwhile, the tender *Cyclamen persicum* on the windowsill introduces relatives that extend way beyond the common and hardy *C. hederifolium* and *C. coum*.

When planting bulbs that expand your repertoire, choose varieties that suit the conditions that are available. Avoid (in the beginning, at least) those delightful but awkward customers that demand extra drainage, the shelter of a glasshouse, or their pot plunging in sand. Instead, seek out new species and cultivars that are perhaps a bit harder to find, but are still easy to grow. Research their preferences and persist; with a bit of know-how and experimentation, any one of these might be a rewarding success or a surprise solution for a tricky spot in the garden.

The choices are extensive: fall-flowering *Sternbergia lutea* loves the sun and bucks the trend for those more accustomed to snowdrops, while nerines like to bake by a warm wall. Colchicums and fall crocuses also like a bright location, while *Galanthus reginae-olgae* and cyclames will take a little shade—as may *Hesperantha*, somewhat surprisingly. *Cyclamen mirabile* makes a dainty feature in fall, given good drainage and humus, while *C. pseudibiricum* adds glamour in early spring.

When it comes to early narcissi, the smaller species, such as cultivars of *N. bulbocodium* and *N. cyclamineus*, tend to arrive ahead of the pack, as does *N.* 'Rijnveldt's Early Sensation.' With crocuses, look for species such as *Crocus chrysanthus* and *C. tomassinianus*, which are convincingly winter-flowering.

Finding out more

There are a number of organizations that count bulbs among their interests, and which often operate internationally. These include the Alpine Garden Society, the North American Rock Garden Society, the Scottish Rock Garden Society (or 'Scottish Rock,' as it is affectionately called), and the Hardy Plant Society. Many of these organizations offer shows, information, and seed exchanges, while some also run tours and expeditions, providing opportunities to see bulbs growing in the wild and to take part in conservation activities. Look for specialist nurseries and local gardens with winter open days, where advice may be on hand and there will be unusual and interesting plants to buy.

Clockwise from top left Pagoda lilies such as *Erythronium revolutum* 'Knightshayes Pink' enjoy a partly shaded site with humus-rich soil. *Narcissus cantabricus* is an early-flowering hoop-petticoat daffodil. Fall-blooming *Sternbergia greuteriana* (pictured) is less hardy than its more common relative, *S. lutea*. Beautiful, airy *Tulipa turkestanica* is a species tulip that flowers in early spring.

The Joy of Snowdrops

Few plants capture the imagination like snowdrops do, and few are as highly anticipated. In the new year they are the first, refreshing sign of a dawning season; an invitation to get outside, to blow the cobwebs away and revel in flowers that are pure, brilliant, and uplifting, and without them, no winter garden is complete.

Known for its charm and as a bringer of hope and light, the genus *Galanthus* is native to a wide area of Europe and parts of Western Asia. Its early history is not well documented, but like many attractive plants it has been adopted and transported as people have moved from place to place, and snowdrops are now extensively naturalized in Britain and Western Europe.

In Britain, the Victorians developed a passion for *Galanthus* that lasted into the early twentieth century. In recent decades, this snowdrop mania has been revived internationally, with new, diverse and exciting cultivars achieving extraordinary prices at auction.

Snowdrops in the garden and landscape

Snowdrops do well in rich, moist, undisturbed soil, taking advantage of the early months of the year to get their cycle of growth in before the canopy closes over, providing cool summer shade and drier soil. They therefore do well in deciduous woodland, but they also flourish along paths and hedgerows, among shrubs, and even in lawns—as long as they get plenty of light and water while they are actively growing and they are not mown back.

Despite traditionally being planted "in the green," meaning dug up and with their leaves on, this risks knocking the plant back if the bulbs and roots dry out. It is often better to plant the dormant bulbs in late summer, and specialist snowdrops are frequently bought growing in pots, where the bulbs will remain moist and healthy.

To create a spread, initially plant out small clumps of snowdrops at irregular intervals, at least 4in (10cm) apart. As the clumps expand, divide every three or four years, replanting to fill any gaps and to extend the spread. The best time to do this is late in the season as the leaves begin to go yellow. For a collection, choose distinctive varieties, plant them at least 12in (30cm) apart, and label well.

In growth, snowdrops are fairly thirsty and they are hungry too; enrich soil with compost or leaf mold as necessary. Feeding will help them to bulk up quickly (see p.143).

On the snowdrop trail

Visiting a winter garden full of snowdrops is one of the highlights of early spring. In Britain, hundreds of gardens fling open their doors to welcome in visitors; and from Europe and Japan to New Zealand and North America, galanthophiles abound.

Theories differ as to how snowdrops came to Britain. Some say it was around the end of the sixteenth century, when they first appeared in the literature, while others claim that they were brought by Norman monks or even that they could have arrived with the Romans.

The truth is lost in time, but with many hundreds of years to naturalize, snowdrops are now widespread and although their glories are fleeting, they have become an essential part of the winter season.

Clockwise from top left *Galanthus* 'Hippolyta.' *G. elwesii* 'Godfrey Owen.' *G. plicatus* subsp. *plicatus* 'Trym.' *G. plicatus* 'Sarah Dumont.' *G. reginae-olgae* 'Naomi Slade.' *G. nivalis* f. *pleniflorus* 'Blewbury Tart.'
Overleaf Massed together, snowdrops have spectacular impact and nowhere is this better illustrated than at Welford Park in Berkshire, UK. Here, an almost unbroken sheet of white carpets the beech wood, and the tall, dark tree trunks rise up among the flowers, like pillars in a cathedral.

The Kitchen Garden

A well-tended vegetable patch is a thing of beauty, yet despite its pleasures, edible gardening is not often considered as a winter pastime. As the first frost cuts a swathe across the landscape, calling a halt to the ripening squashes, blackening the beans, and turning nasturtiums to mush, the mind may wander and the ground lie neglected. Yet there are many practical things still to do.

The soil needs care and nourishment and fruit trees require attention. Hardy crops face the season with fortitude and early sowings herald a summer harvest. Plants may battle on despite the chill, the marigolds blooming bravely in one last hurrah as the seedheads of sunflowers and artichokes enter a new and sculptural phase.

As summer ebbs, other charms come to light. The classic potager reveals its spare elegance; neat ranks of overwintering leeks delight the eye; and trained fruit flaunts its deliciously symmetrical aesthetic. In this spare, lean world, the functional assumes a starring role; the greenhouse becomes an invitingly angular focus, and rhubarb forcers, cloches, and even pea sticks are given an ornamental transformation through the alchemy of nature.

The Veg Patch *in* Winter

Winter harvests typically take time to reach their zenith, but while the process is not a quick one it is worth the wait. The hardiest varieties will stand tall in the coldest weather and can be picked as needed, for seasonal freshness and flavor.

In the modern world, supermarket shelves are always stacked, and with most fruit and veg available all year round, it is easy to become detached from the seasonality of our food. The weather does not always help with motivation either, and while the kitchen garden in summer is a lively and productive place that is full of possibilities, in the depths of winter it feels cold and static.

But although it may seem as if nothing could possibly grow, this is not necessarily the case at all. Without much in the way of international trade and prepacked bounty, our forebears would have grown crops to harvest in winter and plentiful starchy veg to store and consume during the cold months. At the same time, they would have taken the first opportunity to sow seeds for harvest later in the year, and we can do likewise.

So while summer and fall are seasons to pick and preserve, winter brings different gardening and gastronomic pleasures. It is an opportunity to revel in the peppery delights of purple sprouting broccoli and hearty kale, celebrate stored root crops and squashes, and take full advantage of traditional, tasty fare such as rutabaga and turnips in hearty hot pots or roasted with rosemary, sage, or other evergreen herbs. Also bringing pleasure to the palate are bold and fresh cold-weather salads, which are often notably punchy in flavor.

Of the edible plants that crop in winter, the majority are alliums or brassicas—members of the onion and cabbage families—together with perennial herbs and certain resilient salads. In addition, there is a group of honorary stalwarts. Root vegetables and winter squash will store for a long period without deteriorating, while late-ripening apples and pears keep well and even improve, as the flavor develops and intensifies.

To make the most of the season, slow-growing vegetables such as cabbage and leeks should be sown in advance so that when the space starts to become free in late summer they can be planted out. Salad leaves can be started in late summer for winter consumption and while root vegetables lose their leaves, the tubers remain good to eat, with some, such as parsnips, benefiting from the sweetening effect of frost.

Left With its robust, peppery leaves, kale is a handsome addition to the vegetable patch. Covering the plants with insect-proof mesh will also stop birds taking advantage of the harvest.

What *to* Sow *in* Winter

Few crops can be sown in the depths of winter, but garlic and certain onions are a hardy exception, and some legumes can be planted just before or after this point. Getting chiles and tomatoes going early, meanwhile, enables them to steal a march on summer.

Garlic

There is something really exciting about growing your own crop of garlic. It goes into the ground in the most insalubrious weather conditions imaginable, then magically grows, against the odds, to produce a gourmet harvest in summer. It's ideal timing for a plant that requires a fairly long growing season, and a period of cold to encourage formation of the bulbs.

Garlic has two main growth patterns: "hardneck" varieties produce a rigid flowering stem and a single ring of large cloves, while "softneck" garlic has many different-sized cloves in each bulb. The latter is slightly less hardy but stores better. The varieties available to gardeners are usually the traditional, regional crops of places such as Provence, Picardy, and Tuscany, marketed with a new name or described according to their appearance.

Plant individual cloves about 6in (15cm) apart and 1¼–1½in (3-4cm) deep in a sunny spot, preferably where onions have not been recently grown, to avoid diseases such as rust and onion white rot. Cover with soil and water them in. Garlic tends to be trouble-free, but remove competing weeds as necessary and give the plants a drink if there is a dry period in spring, while they are fattening up.

Harvest in summer when the leaves yellow and, in the case of softneck varieties, the stem collapses— don't leave them in the ground or they will rot. The bulbs can be used immediately or dried and then plaited for storage.

Peas and broad beans

Choose hardy varieties, such as broad bean 'Aquadulce Claudia,' which has been bred to withstand cold weather, and wrinkled peas 'Amelioree d'Auvergne,' 'Douce Provence,' and 'Meteor,' which can all be sown in spring while the weather is still cool.

Late winter sowings are fairly reliable, while in mild areas these plants can be sown in late fall as well, to establish before the midwinter chill slows them down. This produces an earlier harvest than spring-sown seeds, although a really hard freeze may still kill them.

Both peas and beans can be sown direct or started off in pots. Peas can also be sown in a length of gutter—ideally, put this in a sheltered location or under cover, and when they are ready to transplant, slide both peas and compost into a preprepared trench. Avoid planting directly into cold, wet soil where the seeds can rot, and watch out for hungry mice, which view peas and beans as a delicious snack. If conditions are good, but no seedlings have appeared after two or three weeks, poke gently around in the pot: if the seeds are not there, plant some more and protect them with a cloche or put them on a table, out of reach of rodents.

When the weather warms, transfer potted plants into open ground, or grow on in a trough or large tub, with a few sticks for support. These plants will perform best in rich, well-nourished conditions.

Top right Broad beans can be sown directly into the soil, as long as it is well-drained and still warm, or started off in pots and then planted out.

Top far right Garlic can be started off in pots or cardboard tubes and planted out later. The many different varieties have subtle and distinct differences in flavor, so it is worth experimenting.

Bottom right Cloches placed over individual plants are useful for keeping the worst of the weather and the pests at bay, and the warm microclimate helps new plantings to get established.

Crops *for* a Winter Harvest

Plants that are established and growing well are less vulnerable to adverse weather conditions, and there are a number of useful crops that will stand through the winter with little consequence. They can then be lifted and used as needed.

Leeks

Delicious, versatile, and easy to grow, leeks can be sown in spring and harvested pretty much all year round. Different varieties mature at different rates and the period of harvesting is usually marked on the packet: 'Below Zero,' 'Winter Giant,' and 'Blue Solaise' or 'Bleu de Solaise,' are good winter choices.

For a winter crop, sow the seeds in late spring, using trays of peat-free potting mix or cardboard tubes with a few seeds in each. When the seedlings are the size of a pencil, they can be transplanted into open ground. Use a dibble or small stick to make a series of holes 6in (15cm) apart, drop a plant into each and water them in. Weed as necessary to keep competition under control and harvest as required.

Leeks tend to need space, and being slow-growing and long-lasting they occupy their section of the vegetable patch for many months. Yet their strong vertical form is an asset, and the varieties that have blue or silvery-green leaves are particularly handsome. If they are forgotten and go to seed, the allium flowers of early summer are attractive, too.

Clockwise from top left Chicory, leeks, chard, and kale are all stalwart winter crops that shrug off a bit of frost and snow. Slow-growing but sturdy, they can be a useful addition to the dinner table for many months, particularly when a number of different varieties are sown in sequence.

Brassicas

Cabbages, kale, and Brussels sprouts are about as seasonal as you can get, and whether served with gravy or shredded in a slaw, they taste even better when local and fresh.

Success with brassicas lies in getting them going in spring and summer; sowing the seeds into trays of potting mix or individual pots and planting them out while the soil is still warm, so they mature as the temperatures dip. There is usually a helpful approximation of the time from planting until harvest on the packet. If you miss the seed-sowing window, or the space is needed for something else earlier in the year, young plants are usually available in garden centers.

In late spring, plant kale such as 'Nero di Toscana' or 'Redbor,' together with cauliflower 'Romanesco,' sprouting broccoli, and Brussels sprouts, all for fall and winter harvesting.

In mid- to late summer, sow oriental greens, mizuna, bok choy, tatsoi, and mustards, as well as chard. These are all smaller than the brassicas above and suitable for pots and raised beds. Turnips, rutabaga, and winter radishes can also be sown at this point and will mature as the cold weather intensifies. If you are a cabbage fan, sow successionally in small batches to keep a constant supply.

Brassicas are a diverse group; specific needs vary and it pays to read the instructions on the seed packet. But as a rule of thumb, beefy, leafy plants need plenty of nourishment and larger, more top-heavy specimens benefit from firm planting so that they do not topple over when mature.

Salads

When pickings elsewhere are thin, salad leaves are fresh, tasty, and easy to grow. They are also compact, making them ideal for pots and small spaces.

Winter salad mixes are convenient, but single-variety packs are also available, including arugula, chard, mizuna, and other Chinese leaves; mustards such as 'Red Lace,' 'Red Dragon,' and 'Green Fire;' and oriental greens.

In addition to lamb's lettuce or corn salad, and delicious pea shoots, there is land cress, which makes a good substitute for watercress; both parsley and cilantro are surprisingly resilient in the face of cold weather, and ferny chervil will usually keep going into the first half of the winter, at least.

Most of these salad crops can be sown to germinate in late summer and mature as the days get cooler and shorter, while hardier varieties can be sown well into fall. Harvest while the leaves are still small and tender, or snip at the seedling stage for microgreens to sprinkle on salads.

Certain lettuces also lend themselves to winter growing and when sown fortnightly from late summer until late fall, and grown on in a cold frame, they can crop continuously until spring. Look for varieties such as 'Valdor,' 'Winter Gem,' and 'Merveille de Quatre Saisons.'

In all cases, sow these crops directly in pots or use trays of good, free-draining peat-free potting mix if you plan to plant out. It is important to make sure that the soil is never waterlogged or frozen. In colder areas, protection from a cloche or polytunnel can be helpful.

Root crops

Technically, most root crops such as carrots and potatoes don't grow in winter, but as storage organs for the flowering part of the plant, their purpose is to last.

Choose varieties that are specifically selected to mature late and keep well; in clay soil these are better lifted and stored, while in light and free-draining soil they can be left in situ until needed. Watch out for diseased or damaged tubers as these may rot, particularly in wet ground. Burrowing slugs can cause damage, as can hungry rodents.

Edible flowers

Many winter flowers should emphatically not be eaten, but there are a few valiant exceptions. Pretty pansies and violas flower through the entire season, while primroses are very early spring arrivals.

Forcing rhubarb

Excluding light from rhubarb in winter encourages the plant into growth prematurely to produce sweet and tender stems that can be harvested in early spring, as a fresh treat when there is little else around.

Key to the process is a mature, healthy, and well-fed rhubarb plant, so feed well the summer before to build up its reserves. When midwinter comes, cover the plant with a rhubarb forcer, bucket, or chimney pot, covering or taping up gaps to exclude all light. Insulating the forcer with straw, burlap, reused bubble wrap, or an old blanket will keep it warm and speed up the process.

After 6-8 weeks the pale pink stems should have reached 8–12in (20–30cm) in length and can be gently pulled from the base of the plant, which should then be left unpicked in summer and not forced the following year, as it will need time to recover.

Exercise caution with standard bedding plants, as they may have been sprayed with chemicals unfit for consumption—it is best to buy organic or grow your own from seed.

Better still, look in the garden or polytunnel and see what is hanging on. Marigolds often last well into the winter in a sheltered spot, geraniums and roses often produce an optimistic late bloom or two, while peppery yellow arugula flowers may not yet have set seed.

Clockwise from top left Rhubarb starts growing early in the year. Homegrown celeriac can make a warming winter soup. Beets will store well when picked and can also be left in situ on dry ground as long as the winter is not too severe. Brussels sprouts are a classic of the season and the crown of the plant can be used as a substitute for cabbage.

The Winter Orchard

Winter is the perfect time to plant dormant fruit trees, which will then crop for many years to come. There are hundreds of different and delicious varieties available and a range of rootstocks means that there is a fruit tree for every situation, from a substantial heritage orchard to a tiny garden, with a couple of dwarf specimens in a patio pot.

In many fruit, the selected variety is conserved by grafting the top wood onto a rootstock. The ultimate height of the resulting tree will then depend on whether the rootstock is standard, which is used for big trees, semidwarfing, or dwarfing. This means that each cultivar can be available in a range of sizes.

Pruning

The pruning of fruit trees is carried out in order to improve their shape, health, and productivity. In apples and pears, this can be undertaken from the point when the leaves fall until the buds break the following spring. Stone fruit such as cherries and plums should be pruned in summer, while fan-trained or espalier apples and pears will need pruning in summer as well as in winter.

Formative pruning of young trees creates a balanced plant with well-spaced branches, while maintenance pruning of older trees removes old growth and stimulates new growth, thus creating a constant supply of productive two-year-old wood. Start by removing any wood which is dead, diseased,

---

Left Espalier fruit trees have a wonderful winter structure and take up very little space.

or crossing, then stand back and have a look at the tree. Open up the center by thinning or removing branches that are facing inward, and shorten long, whippy new growth by about a third.

Old trees can be renovation-pruned to reinvigorate them and make them more productive, but take care with venerable specimens as cutting off too much wood may irrevocably damage or kill them. An old fruit tree has considerable environmental and wildlife value, so it can be best to leave well alone and enjoy the rugged structure growing as nature dictates.

Pests and diseases to watch out for

Brown rot is a widespread fungal disease where tree fruit develops brown patches with rings of creamy pustules. Dispose of affected fruit in the green waste or by burying at least 12in (30cm) below the soil's surface, rather than composting. Thinning the fruits can help, but to a great extent, this is something that fruit growers must learn how to live with.

Fire blight is a widespread and destructive disease which makes leaves look as if they have been scorched. Caused by the bacteria *Erwinia amylovora*, it is particularly common in southern areas of Europe and North America. Act fast to prune out affected branches or remove trees and then burn the infected material.

Apple canker is the result of infection by the fungus *Neonectria ditissima*, which causes areas of dead, sunken bark and can also cause shoots and branches to die back. Some cultivars are more susceptible than others. The treatment is to prune out and burn affected wood; canker rarely kills the apple or pear tree, which may recover to a certain extent.

Codling moth is the cause of "maggoty apples"— and other fruit, too. Caterpillars of the moth *Cydia pomonella* bore into the fruit, which may then drop early. Encouraging natural predators is preferable to using pesticides.

Seasonal Soil Care

Whether plants are grown in open ground, a greenhouse border, or large containers, they need a good, free-draining soil and plenty of nourishment if they are to thrive and produce a decent harvest.

In the vegetable patch, lightly cultivate the soil in fall, removing competing perennial weeds and adding organic matter if necessary. Traditionally, kitchen gardens were heavily worked and dug over regularly, but this is now widely thought to be unnecessarily disruptive, with impacts on soil health, its microfauna, and desirable attributes such as drainage.

Before cold weather sets in, spread a good layer of organic mulch over the surface of the soil to improve its condition and suppress weeds. Spent potting mix, manure, leaf mold, and composted bark are all ideal, and will be gradually incorporated into the soil through the action of invertebrates and weather.

If open ground for growing vegetables is in short supply, many plants can be successfully grown in containers all year round. The pot must have drainage holes and be large enough to accommodate the roots and the full-grown plant, while the growing medium should be fresh and rich in nutrients.

Right Protect and feed the soil with a good layer of organic mulch—homemade compost is ideal. Spread liberally on top of empty vegetable beds and around established trees and bushes; this improves soil structure and adds the raw materials that a healthy community of soil organisms needs to thrive.

A Homegrown Harvest

Growing new plants is part of the gardening cycle, and taking cuttings, dividing clumps, and collecting seeds for crops in the future is a pleasurable pastime even in winter.

Keeping the edibles coming for as long as possible makes perfect sense, yet we must work with the rhythm of the land. As fall closes the year, seeds can be saved and stored, but winter is fleeting, and as the days draw out, crops can be propagated and new sowings made with the fresh, lush plants of a new season in mind.

Propagate perennial herbs

Like certain salad leaves, short-lived herbs such as chervil and parsley can be kept going in all but the coldest months with some protection.

Winter is also a good time to propagate and divide other herbs too, particularly established clumps of mint, chives, garlic chives, marjoram, and oregano. Dig up the plant or expose the root ball, and use an old bread knife to cut it into sections. Replant a section where the mother plant was, and spread the rest around the garden or pot them up to give away.

In late winter, bringing the potted herbs under cover will encourage them to sprout fresh green leaves for an early harvest.

Far left Mature bean pods will dry naturally on the plant, ready to be harvested for later use.

Left Chiles can take a long time to get going, but brought indoors they will fruit well into winter, as shown here alongside stored squashes and forced narcissi.

Overleaf In a vegetable garden, cloches and rhubarb forcers are both useful and ornamental.

Saving seeds

It is simple to save seeds—just collect them when they are ready and dry them. Open-pollinated older varieties are usually a better choice than modern F1 seed, where the resulting plant is likely to be less good than the highly bred parent.

Take the seeds from a ripe fruit, seedpod, or seedhead. If it is a pulpy fruit, such as a tomato, scrape off the jelly as it will inhibit germination, then allow the seeds to dry before storage. Chiles, beans, and corn can be dried in an airy place prior to removing the seeds.

Many tree nuts can be planted immediately, as can salads and herbs. With annual crops such as beans and squashes, store the ripe seeds in a paper envelope in a cool, dry place, to sow in spring. They probably won't come true, but part of the fun is in growing something new, interesting, and hopefully tasty.

Looking ahead to summer

Although crops such as chiles, eggplants, and tomatoes are not winter vegetables, they need time to grow and a long, warm season for the fruit to ripen, so where their seedling needs can be met, they benefit from an early start.

Slow-growing and fairly compact, chiles can be sown a few seeds to a pot in late winter. Cover with a little potting mix, water well and put the pot in a warm, bright, and humid location—inside a heated propagator on a windowsill or in a conservatory is ideal. Pot on when the seedlings are large enough to handle. In spring, move the young plants into a greenhouse if available, or keep as houseplants until early summer when they can go outside.

Tomatoes and eggplants can be sown in the same manner, but tomatoes in particular are large, fast-growing plants. The seedlings need bright, warm conditions to grow—the temperature should not drop below around 59°F (15°C)—and if sown indoors too early they can become pale and leggy, so where there is no greenhouse it is better to wait until spring.

The Cutting Garden

While the weather in winter can be bright, crisp, and glorious, this is also the season of torrential rain, howling gales, and periods of sullen drizzle. For every day of blue sky and bright sunshine, there will be others less disposed to make the heart sing.

But there are plenty of ways to enjoy the garden indoors in comfort and warmth, and while the sweet peas of summer may be a distant memory, a vase of loveliness at this time of year is still possible.

Year-round color in florist shop windows has dulled our sense of seasonality, with plants grown under glass or shipped from warmer climes, but awareness is growing and a new trend is emerging: one that cherishes the sustainable fruits of our gardens and of small producers, valuing enterprise and rejoicing in the possibilities of locally grown arrangements.

Even a small garden can offer colorful stems and tiny, fragrant flowers; berries, seedheads, and skeletons. Twigs can be slender and elegant, or knobbly and lichen-encrusted. The seasonal dishabille may reveal the bark of seemingly ordinary trees and shrubs to be textured and intriguing, while evergreens are bright, bold, and inspiring.

A Foraged Floral Harvest

Winter flowers are few, yet those that do exist are delightful. Dainty, fragrant blooms appear on nude stems, waiting to be plucked and combined with berries and seedheads, foliage and grasses, to create something that is both special and unique.

A winter cutting garden is a lovely thing to plan for, but when spontaneous floristry is required, many gardens are much more well-supplied than they may at first appear. Going outside on a fresh, bright morning and just looking, can reap dividends. Shelve any preconceptions as to what floristry is, or what a bunch of flowers should look like, and keep an open mind. A handful of early daffodils might be a bonus, but this is far more a case of seeing what there is in terms of form, texture, and detail, and letting the creative juices flow.

By now, seedheads are skeletal—gently desiccated by chill winds and advancing age, yet still possessed of faded elegance aplenty. The bold, whorled flowers of *Phlomis* mature to crisp ruffs arrayed along a tall stem, while the beefier of the umbellifers retain their structural integrity, and a waning tithe of seeds, for months. Poppies are stately in decay and alliums petrify, like fireworks frozen in the moment before they vanish into dust and darkness.

There is joy to be had in the ephemerality of plant material that has a brief moment of glory and is gone. When conventional blooms are few, it pays to capitalize on the blink-and-you'll-miss-it flare of grasses, with their sleek, burnished stems and airy seedheads. Watch clematis flowers segue into punky spiders or shocks of soft fluff, and honesty in its final moments before the silvery moon-pennies disintegrate, leaving only a rim and a shadow.

Even later in the season, there are still plenty of options. Twigs and greenery are staples at this point, but plants that have a flowerhead made up of robust, leafy bracts are particularly useful; they demonstrate considerable tenacity, neither withering nor flopping.

In a sheltered spot, hydrangeas can last all winter, their flowers changing color as they age. Fresh pastel hues fade to green, teal, or plum, then finally a translucent taupe that, when backlit by a shaft of sunshine, flares caramel and gold. Eryngiums and teasels (considered invasive in some parts of the US), meanwhile, have distinctively spiky flowers that resemble medieval weaponry, and they will stand proud for months, waiting to be picked.

Left Elegant twigs, ethereal seedheads, and bright berries are all handsome additions to a winter bouquet.

Winter Gems

Winter brings Christmas and the holidays; halls are decked and logs crackle on the fire; the sharp scent of pine is in the air and glad voices sing *Gloria in excelsis Deo*. At this point, tradition dictates that the garden gives generously, with swags of ivy and bunches of holly. This may be the cultural recollection of Christmas and holidays past, and centuries old besides, but the present-day flower arranger still has a multitude of choices.

A quietly lovely and ultralocal selection of plant material needs no fanfare of trumpets. Planting a cutting garden with winter in mind, or simply foraging an arrangement from the garden that already exists, provides a display that is interesting, impactful, and unfussy, and an opportunity to appreciate the stems, flowers, and foliage up close.

Harvested for the house, winter-garden staples can be repurposed and repositioned, allowing them to perform in a new way. It is no longer about symbolic greenery and the brief, torrid flurry of a religious festival, or even about garden design or horticulture. Instead, it is something quieter, more personal and more poetic. A meditation on decoration and ornament, and the satisfaction of creating something new from found things; the opportunity to celebrate personal goals and anniversaries with posies or gifts; or to stretch existing knowledge of floristry and take it to exciting new places.

The spectacular stems of cornus and willow can be cut elegantly long and displayed in a statement vase, or woven into a frame as the basis for a wreath or dried-flower arrangement. Other woody specimens offer up exciting possibilities, too. Fine, lax birch twigs form swags of purple and chocolate.

Hazel twigs, meanwhile, are subtle greenish-gray when young and become silver-brown as the tree matures, and the cut branches are adorned with gently expanding catkins.

With plant material scarce, less is more. Form becomes everything and structure, shape, and poise are key. Here the interest can come from an unexpected quarter: the plants with thorns, intriguing kinks or textured bark, or fallen branches festooned with lichen that become an artwork in their own right.

Cones and seedpods also provide a compelling sense of texture and rhythm. European larch drops twigs that are encrusted with neat cinnamon cones, while alder, which is often overlooked, has small but delightfully sculptural black cones that remain on the tree well after the seeds have dispersed.

It is a moment to appreciate the present: at no other time of year does pussy willow hold such promise; the summer night never carries its perfumes with such an element of surprise; the hope that is almost tangible in the buds that change color and get larger daily, as they yearn for spring and anticipate the sap that will soon rise and catapult the plants headlong into leaf and bloom.

With so many secrets waiting to be discovered, even the winter tidy-up can be profitable, and when cutting back and pruning, take a fresh look at what would usually go in the compost bin or shredder. Knobbly pear and apple twigs are charming in a small vase with a few early daffodils and a sprig or two of ivy, while the spent blooms of hydrangeas, *Verbena bonariensis*, and *Hylotelephium spectabile* will make an on-trend dried arrangement.

Clockwise from top left The florets of *Hydrangea macrophylla* 'Altona' age gracefully. Curly hazel, *Corylus avellana* 'Contorta,' has twigs that are delightfully erratic. Many winter flowers, such as *Sarcococca hookeriana* var. *digna*, have a scent that surprisingly belies their size. Long-lasting larch cones will add structure and punctuation to an arrangement.

Bulbs *in* Bloom

Bulbs usually make excellent cut flowers, but while daffodils, tulips, and ranunculus appear in every florist and supermarket, the choice is limited by the demands of the floristry trade. Yet bulbs are easy to grow at home and planting some for cutting purposes, or simply gleaning a gentle portion from what already exists, gives a much more varied and sophisticated display, along with a better connection to the natural world.

When it comes to cut flowers, the commercially successful lines have been bred to have long stems and large, uniform blooms. They have to be long-lasting enough to withstand the rigors of transportation, and early in the season, these are likely to have been brought on under cover and then imported.

But rather than settling for the homogeneity of retail, picking flowers as they emerge in the garden gives a sense of living in the moment, and offers a far greater choice. As blooms emerge, they can be plucked and enjoyed, while options and opportunity can be maximized by growing earlier- or later-flowering forms, which will also spread out the peak of the season. Access to a cold greenhouse

(see p.182), or even using a large pot to cultivate bulbs in a sheltered spot, means precocious flowering can be encouraged at home.

Snowdrops are rarely used as a cut flower for the simple reason that tiny, dainty blooms with an expensive reputation are not commercially viable. Yet they cut beautifully and look gorgeous arranged in a container that is in proportion to their size. The elegant, often scented flowers are immensely variable (see p.146–147), so taller ones can be teamed with twigs or foliage, and a simple mixed posy can be a study of the diversity of form and the purity of green and white.

A number of other bulbs are almost entirely off the radar as cut flowers, too. Snowflakes, or *Leucojum*, are not usually considered for indoor displays, but given their height and elegance this is a major oversight. While they may be small, the flowers of glorious, vibrant blue *Scilla siberica* make a superbly eye-catching buttonhole, while *Eranthis*, the glossy, golden winter aconites, will grace a tiny container. Ridiculously easy to grow, *Muscari* can be a minor menace in the garden but its pleasing pink, white, or blue flowers cut well, as do those of *Chionodoxa* and *Iris unguicularis.*

Some bulbs don't really lend themselves to cutting, but with a bit of imagination they can make a very attractive indoor display, nevertheless. For example, soft-stemmed crocuses and *Iris reticulata* can be grown in vases that are similar to those used for hyacinth bulbs, but smaller. With a greenhouse or polytunnel, meanwhile, it is fairly easy to grow certain commercial varieties for the purposes of cutting at home. These could include paperwhite narcissi, brightly colored ranunculus, cultivars of *Anemone coronaria*, and even amaryllis, started off early under cover and then cut for the house.

Left Many bulbs are suitable for cutting, and beautiful, fragrant paperwhite narcissi can be forced to bloom from midwinter, when they can be cut for a vase. When members of the *Narcissus* family are cut, they exude a mucilage (or slime) from the end of the stem and florists find that this often disagrees with other plants when they share a vase. To get around this, enjoy daffodils and paperwhites by themselves, perhaps with the addition of robust foliage. Alternatively, leave them for a couple of hours to stop oozing; the stems can then be used to create a mixed arrangement without issue.

Winter Flowers

The flowers of the season may be few and they are frequently small, but they are lovely, and many are also strongly scented, which adds a powerful and unexpected new dimension to an arrangement.

Stars of the season

Narcissi and even snowdrops are well known for their fragrance, but it is the winter-flowering shrubs that are the star performers in this area. Just a few sprigs of *Sarcococca* or *Lonicera fragrantissima* will scent a room, while *Chimonanthus praecox* earns its common name of wintersweet.

Daphnes pack an olfactory punch, and *Daphne bholua* 'Jacqueline Postill' is unmissable when it is in full spate. The gentle bubblegum-pink flowers of *Viburnum × bodnantense* 'Dawn' appear on and off all winter, bouncing back after every hard frost with new clusters of bloom, but for nostalgia, the classic winter-fragrant all-rounder is *Mahonia*, with its exciting, spiky leaves and racemes of yellow flowers.

Diverse blooms

In sheltered conditions, many late summer and fall flowers such as chrysanthemums, *Vernonia crinita*, and *Serratula tinctoria* var. *seoanei* will carry on well into early winter too, but other flowers are poised to steal the show with their element of surprise. The branches of *Hamamelis* sprout crinkled, fragrant, citrus tassels, and *Prunus × subhirtella* 'Autumnalis' produces delicate cherry-blossom flowers from late fall to early spring. *Salix gracilistyla* 'Mount Aso' has fluffy, implausibly pink, pussy-willow catkins, while *Jasminum nudiflorum* is known in China as *Yingchun*, the flower that welcomes spring, for its buttery-gold flowers that burst from the leafless stems.

There are also a number of winter-flowering evergreens, including mahonia, camellias, and rhododendrons, and these produce relatively large flowers for the season. It is worth seeking out *Rhododendron* 'Christmas Cheer'—although it usually flowers rather later than its name implies—and *R. dauricum* 'Midwinter' AGM.

Not all glamorous winter flowers are good for cutting, however. In a vase, camellias are beautiful but ephemeral, while fresh hellebores, though excessively tempting, will quickly wilt unless they are cut after the seeds start to form and the petals— or, properly, sepals—have had a chance to harden.

Forcing cut stems

If spring can't come quickly enough, stems of early-flowering shrubs and trees can be brought inside.

- Pussy willows and hazel catkins are quick to respond to the increase in temperature and buds will rapidly start to expand.
- *Ribes*, *Amelanchier*, and *Chaenomeles*, or Japanese quince, are usually pruned after flowering, but cutting a few stems for early flowers gives a longer period of enjoyment and a head start on spring jobs.
- When pruning apple and pear trees, select stems that have plenty of fat flower buds and put them into a bucket of water. Peaches and apricots flower earlier still, and are beautiful.
- Toward the end of the season, experiment with less usual subjects such as magnolia, horse chestnut, *Philadelphus*, or even *Cercis siliquastrum*.

To force flowers, cut the base of each stem at an angle of 45 degrees and submerge it immediately in a heavy container half-filled with water. Keep in a cool place, away from radiators and direct sunlight. Misting regularly will help to maintain humidity and encourage the flowers to expand. Surplus stems can be left in a shed or garage, in a bucket of water, and brought out over a period of weeks to encourage into bloom as required.

Clockwise from top left The pink, fluffy catkins of *Salix gracilistyla* 'Mount Aso.' *Daphne bholua* 'Jacqueline Postill' has gloriously fragrant blooms. *Viburnum bodnantense* 'Dawn' flowers for months.

Foliage, Stems, and Fruit

Stems and foliage are the backbone of a winter garden and they lend themselves to seasonal flower arrangements as well. The subtleties of form and spare twiggy structure can be charming, yet some plants pull out all the stops with bright colors and exciting berries.

Foliage

Holly and ivy may crown the season, but it pays to look beyond ordinary dark green to the wide range of ornamental cultivars that are available, with variegated leaves or yellow holly berries. Conifers are also highly diverse and can contribute a range of qualities, including bushy, prickly, or drooping branches, that look great in arrangements.

Variegation is not to everyone's taste, but it does create a lively effect and splashes of gold and cream come into their own during the darker days, lightening the gloom. *Osmanthus heterophyllus* has hollylike leaves that are stippled with sulphur yellow, and while *Aucuba* and *Euonymus* can seem brash in summer, in a more muted context they are striking rather than gaudy.

Some plants seem to go out of their way to be conspicuous in winter. Eucalyptus is much sought after by florists, while *Nandina domestica* has acid yellow or reddish foliage, and *Pittosporum* is fantastically varied and versatile in arrangements, with leaves that range from pink and purple to peppermint cream. What's more, some evergreen shrubs have multiple positive attributes; the berries on holly trees are hotly anticipated, but the sharply sculptural dead flowers of *Escallonia* and *Abelia*, though diminutive, can be quite startling.

Stems and branches

An armful of bright cornus is bold and fabulous in a heavy jug, but there are other subtly lovely candidates for cutting too, such as birch with its tones of coffee or plum; maple prunings, which may be intriguingly striated; or even the tangerine shoots and crimson buds of *Tilia cordata* 'Winter Orange,' which is not often considered, but has plenty of panache.

Some slow-growing woody plants grow attractively gnarly with age, accumulating character, moss, and lichen over time and turning even a twig into a fascinating miniature landscape. Oak is a fine example of this (as is hawthorn, for the less superstitious), while knobbly larch twigs are encrusted with cones.

In a larger garden, curly hazel (*Corylus avellana* 'Contorta') and corkscrew willow (*Salix babylonica* var. *pekinensis* 'Tortuosa') are pleasant enough in summer, but very striking in winter, and they are both substantial and prolific enough to please even the greediest of florists. Some hedgerow trees such as ash, meanwhile, can present a sophisticated palette of gray stems and black buds, while elegant young beech twigs often hang onto their dried leaves until the new foliage appears.

Fruit

Many fall berries are remarkably persistent and look utterly lovely as single stems or as part of a mixed arrangement. Look for crimson rosehips and the smart black berries of ivy to add to a winter bouquet, perhaps with long-lasting crab apples.

Berberis and viburnum also have interesting, characterful fruit, if the hungry birds don't eat it first, and with a bit of forward planning, the display can be augmented with oddities such as the neon orange-and-pink seedpods of *Euonymus europaeus* and the improbable purple caviar of *Callicarpa bodinieri*.

Clockwise from top left The fruit of *Malus* 'Comtesse de Paris.' Willow stems are extraordinarily varied in hue. *Viburnum davidii* has pewter-blue berries in winter. In addition to good leaf color, *Nandina domestica* produces cascades of red berries.

Dried Flowers

After half a century in the doldrums, dried flowers are back in vogue, shaking off the brown and crispy aesthetic that was all the rage in the 1970s and strutting their stuff anew. Long-lasting arrangements, packed with style and full of inventiveness, are a refreshing alternative to the seasonal norm, as well as being an environmentally friendly option.

Summer flowers for winter use

Collected from the garden when aged and desiccated to perfection, or harvested at their peak to preserve color and structure intact, dried flowers are elegant in sophisticated contemporary designs. Rather than using—and regularly replacing—fresh flowers, dried blooms can be an economical and less resource-intensive approach, and when they are locally grown, they are more sustainable still.

While some flowers and seedheads hang on robustly into the fall and winter, many will rapidly deteriorate in the face of heavy rain and battering winds. However, picking them in advance and drying them means that the plant material will remain in good condition and can be used for wreaths and arrangements at a later date.

In this way, not only will the soft seedheads of shrubs such as *Cotinus* retain their structure, but summer flowers such as delphiniums and rudbeckia can be enjoyed long after their season has passed.

"Everlasting flowers" are also seeing a resurgence in popularity, and in a sunny spot it is worth experimenting with annuals such as *Helichrysum bracteatum*, *Helipterum roseum*, or *Xeranthemum annuum*. Not only do these cut and dry extremely well, they retain their bright colors too.

How to dry flowers and seedheads

There are various ways of drying plant material. Some methods involve desiccating agents, such as silica gel, but the easiest way is to dry flowers and seedheads just as if you were drying a bunch of herbs.

Collect seedheads at their peak and cut flowers on long stems, just before they open fully, as they will expand further as they dry. After removing most or all of the leaves from the main stems, divide them into small bunches using elastic bands. Larger, moister blooms such as roses, lilacs, and hydrangeas should be dried individually. Suspend the seedheads and flowers upside down from a coat hanger and leave them in a cool, dry, and airy place for at least a fortnight.

Best results are achieved by using naturally thin, dry or papery flowers as these dehydrate quickly and often retain their color well. Honesty (*Lunaria annua*), hops, lavender, and hydrangeas are classics, and astrantia, statice, and gypsophila also work well. Annual flowers are popular, too; a spring sowing of cornflowers, nigella, or marigolds can yield a long-lasting harvest, and even common wild plants such as buttercups and clover are pretty when dried.

With seedheads, go for strong shapes and interesting textures. Stems of wheat and bunny tail grass (*Lagurus ovatus*) are rather overused, but it is hard to go wrong with structural opium poppies and *Scabiosa stellata* 'Ping Pong.' Other grasses that are effective include the various forms of *Miscanthus*, *Anemanthele lessoniana* (pheasant tail grass), *Calamagrostis brachytricha*, and delicate *Agrostis nebulosa*.

Dried foliage is as useful as the fresh form, so it is worth experimenting with a range of leaves together with berries such as rowan or rosehips.

Once the plant material is ready, it can be arranged in a vase or used in a range of other floral creations. Avoid damp locations and keep out of direct sunlight, which can cause the colors to fade.

Top far left Honesty (*Lunaria annua*) dries to form pretty moon-pennies.

Top left *Eryngium* is particularly striking in frost.

Bottom left In hydrangeas, the summer pastel colors deepen as the flower ages, finally fading to brown.

In the Greenhouse

Unheated greenhouses are often the exclusive domain of the kitchen gardener, but when the vegetables are gone (or, at least, pared back to winter salads), the comparative warmth and shelter can also be used for extending the season and improving conditions for homegrown flowers. It is also a congenial place for the winter gardener to get on with sundry small jobs on a sunny day.

A frost-free greenhouse or conservatory helps keep fall-flowering plants going for longer and get spring-flowering plants going sooner. It is the ideal space to start off seeds for both home-grown veg and cut flowers, and to overwinter tender plants and windowsill pelargoniums, too.

When protected by glass, chrysanthemums can reach their natural conclusion in early winter, overlapping with paperwhite narcissi planted a couple of months earlier. *Amaryllis* also get off to a sturdy start in a polytunnel, before being brought inside to flower, while citrus trees, while not usually kept for cutting, produce sweet-smelling flowers in winter and are attractive candidates for a cool conservatory.

Where there is a vacant border in a polytunnel or greenhouse, filling it with bulbs in late summer will bring forward their flowering time for locally grown, early-season flowers such as daffodils and tulips. In commercial circles, ranunculus and anemones are floristry staples in spring and winter, and it is fairly straightforward to grow these cool-season flowers at home under cover, as long as they are given excellent drainage and as much light as possible.

For a committed home florist, a greenhouse or polytunnel is a huge asset, enabling hardy annuals such as *Ammi majus*, *Cerinthe*, snapdragons, and sweet peas to be sown sequentially from fall through to spring. Where there is space, leaving the first few sowings inside to flower, and planting later ones outside when the weather warms, extends the picking season still further. At the same time, half-hardy annuals such as *Cleome*, *Amaranthus*, and *Moluccella* may be started off earlier than would be possible outdoors, without the inconvenience of cluttering up every windowsill in the house.

While it is important not to present tender plants with entirely inhospitable conditions, many seeds will germinate under cover in early spring without having to use a heated propagator. They won't be as quick off the mark, but equally, they won't need to acclimatize to cooler conditions in order to grow on.

One of the greatest advantages to having a bright and covered space, however, is its potential for overwintering some of the fussier plants of the late summer garden. Dahlias, cannas, and ginger lilies are fast-growing and fabulous when late spring comes, but they like to be dry and frost-free over winter. So, whether they are lifted out of heavy soil and stored in boxes of dry compost, or are brought inside, pots and all, the protection helps. When they are ready, they will start to shoot, and being able to maintain a greater level of control over the surroundings makes decimation by slugs and snails far less of problem.

A cold greenhouse is also a good place to get forced stems started (see p.182), for a taste of what is to come. Kept cool and bright, the buds on cut branches of blossom will gradually swell, and can be brought into the house and enjoyed as needed.

Right Whether used for sowing an early batch of seeds, overwintering tender pots, or forcing cut stems, the greenhouse is an asset in the winter season.

Winter Celebrations

Though fulsomely plundered for Christmas wreaths and holiday decorations, the garden is not yet done with rejoicing. For every event and celebration the season brings, there will be a fresh and fragrant posy too.

Life does not stop in winter; indeed, the connection to the earth and the weather can be more visceral than ever. But when it comes to cut stems and seasonal arrangements, we are distracted by what is available commercially, and often forget about all the things that the garden has to offer. By looking closer to home, however, a gift or decoration can be personal and thoughtful in a way that supermarket blooms can never be—and not only is it easier on the pocket, it is also a greener option.

Flowers are a meaningful part of many life events. They play a role in winter festivals such as Candlemas, when churches are decorated with snowdrops, and Mother's Day is often recognized with a bunch of blooms. But the commercial big hitter of the season is Valentine's Day, where dozens of out-of-season red roses, lavished with chemicals and flown across continents, are bought at great expense in an ironically unsustainable token of enduring love.

Yet there is nothing romantic about polluting a struggling Earth with ostentatious gestures, and there is another way. A bunch of flowers foraged from the garden, wrapped in natural ribbon or twine, travels feet not miles and does away entirely with plastic wrapping. It is a gift that speaks far more surely of regard for a mother, a lover, or a friend. Humble it may be, but with something local, fresh, and homegrown, you are also giving them the world.

Right A fragrant posy of witch hazel (*Hamamelis* × *intermedia*), cut and placed into a simple terracotta pot.

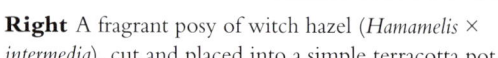

Bright Stems and Evergreen Foliage

In this, the most muted of seasons, some plants remain defiant. Undiminished by wind and weather, evergreens provide a comforting element of permanence. And although a few are steeped in tradition and others have fallen from grace, there is much more to cold-weather foliage than might first be imagined.

For a winter garden that is impactful, exciting, and fun, taking the concept of an evergreen to its logical extremes makes sense. All the available shrubs and trees must be considered: from specimen plants to hedging and climbers, low-growing groundcover such as bergenias and heathers, and with an honorary category reserved for cornus and willow, which slough off their summer shroud to reveal stems of unparalleled brilliance.

When most leaves are gone, "evergreen" is the default term for what is left, but it is reductive, even misleading. These plants are a constant and colorful presence, persisting as their herbaceous and deciduous companions rise and fall. Green is an option among an array of yellow, orange, purple, and red, spotted and splashed, and stems and foliage are an enormously versatile element of the winter garden.

A Fresh Look at Evergreens

Underappreciated and often disparaged, evergreens usually miss out on the fanfare that greets perennials and bulbs, but all that is about to change. A new interest in winter gardening is driving their renaissance and these colorful, leafy cold-season plants are once again becoming popular.

Evergreen plants can be easy to ignore. Tough as old boots, they are so common that they might as well be invisible, but rugged characters such as pyracantha, *Choisya ternata*, *Viburnum davidii*, cotoneaster, euonymus, and even hebe and privet hold their own, and they live on in our gardens regardless.

Whether planted with a purpose or a haphazard hangover from a former time, existing evergreens are a good winter starting point, providing a useful splash of color and a backdrop to more focused displays. Easy, seasonal foliage plants such as holly, ivy, skimmia, and heathers may be marketed as winter interest, yet as a group, evergreens are not universally popular, often beset by image problems, pests, or diseases.

Once a much-loved garden stalwart, box (*Buxus*) now suffers from a range of challenges, including pernicious box blight and box tree moth (see p.198) which, in some areas, render the plants unusable, although solutions and alternatives are being sought. And while yew is highly acceptable, other conifers are frequently considered dated or unruly.

Leylandii cypress is often planted as a fast-growing hedge but it will swamp a smaller garden, and has given the whole tribe a bad name. The dwarf conifers planted in the 1970s, meanwhile, were declared twee and unfashionable soon afterward. This slur not only stuck, but was rolled out to cover anything that vaguely resembled an ornamental pine, although there are signs that the tide is now, finally, turning in their favor.

Variegated shrubs also divide opinion, but there are suggestions of a comeback. Unpopular due to their brashness and a heavy association with mid-century shrubberies, aucuba and euonymus are now finding redemption with the increased interest in home flower arranging (see p.178).

It is, therefore, not really surprising that people look askance at traditional evergreens, focusing instead on plants such as phormiums, which are thought of as low-maintenance and less problematic.

But with winter in mind, it is worth taking a fresh look at what is already in the garden. A carpet of heathers might offset a handsome specimen tree and pyracantha can be pruned to improve its behavior, while large shrubs and many conifers can be managed to perform in a more useful and sophisticated way (see p.190 and p.196).

Left Drifts of heather and a range of handsome conifers create a long-lasting and interesting display.

Using Conifers

Often generically called pine or fir trees, conifers are a diverse group of gymnosperms that predate flowering plants by several hundred million years. Their mostly evergreen and structural nature has year-round garden value, but they are often poorly utilized.

Found almost all over the globe, and made up of a wide variety of cone-bearing plants, conifers range from timber trees and spectacular landscape specimens to dwarf and alpine shrubs. The majority of these are evergreen, although deciduous species do exist, and in the winter garden, conifers are a force.

Unbeatable in their hardiness, interest, and range of color, and for the structure they bring, evergreen species include *Pinus*, *Abies*, *Podocarpus*, *Cupressus*, *Picea*, *Juniperus*, *Cryptomeria*, *Thuja*, *Cedrus*, and *Taxus*, while deciduous conifers include larch (*Larix* species) and dawn redwood, *Metasequoia glyptostroboides*. Among these, habitat preferences vary: some like dry, well-drained conditions, while others will grow in fairly heavy soil; some like sun, others are far less fussy. There is a wide range of sizes, shapes, and colors, too.

As a result, there are conifers for every garden, and they have considerable design value. Narrow, pointed forms can act as an accent or exclamation mark, while a repeated series of bright yellow pines will draw the eye.

Where once a box ball might have grown, a compact globe of *Cryptomeria* will do the same job. Different shades of green, blue, gold, bronze, and even purple can be layered to great effect and the cones of many species are handsome as well.

Selecting conifers

Conifers come in all shapes and sizes—tall, short, spreading, and sculptural—but when choosing a plant for color or form it is also worth picking something that will enjoy the available growing conditions too. Fortunately, with such variety, it is highly likely that there will be a tree to suit any given situation.

Tall and thin: the slender cypresses of an Italian landscape are beguiling, and in a well-drained and sunny location you can get the look with a tall *Cupressus sempervirens* Stricta Group or *Chamaecyparis lawsoniana* 'Columnaris.' For shadier spot or where the soil is more moist, consider a columnar yew such as *Taxus baccata* 'Fastigiata,' while pencil-thin *Juniperus scopulorum* 'Skyrocket' won't outgrow a small garden.

Yellow and bronze foliage: with a bold effect in a small package, *Pinus mugo* 'Winter Gold' lives up to its name, while slightly larger *Thuja plicata* 'Stoneham Gold' and beefy *Cupressus macrocarpa* 'Goldcrest' are more chartreuse. For orange-bronze foliage that changes with the season, try softly cascading *Cryptomeria japonica* 'Elegans' for a medium-sized tree, or small and rounded *Podocarpus* 'County Park Fire' for something more manageable.

Fine form: in landscapes and larger gardens, it pays to plant evergreen trees to be enjoyed for generations to come. The monkey puzzle tree *Araucaria araucana* is always an exciting specimen, and cedars such as *Cedrus atlantica* Glauca Group and *Cedrus libani* are magnificent. Scots pines, *Pinus sylvestris*, become more sculptural with age while deciduous *Metasequoia glyptostroboides* looks fantastic silhouetted against the sky. All of these are around 38ft (12m) tall, although spread may vary.

Smaller and rounder conifers: For many years, box (*Buxus*) has been the go-to for clipped evergreen balls, but where it now struggles, check out feathery *Cryptomeria japonica* 'Elegans Compacta,' glaucous *Juniperus squamata* 'Blue Star,' *Picea pungens* (Glauca Group) 'Globosa,' or dense *Thuja occidentalis* 'Little Gem,' all of which have a naturally curvaceous form and are not that large, in the scheme of things.

Right In a dry garden filled with Mediterranean-style planting, Italian cypresses (*Cupressus sempervirens* Stricta Group) are very much in keeping.

Brilliant Stem Color

When gales whisk away the last leaves of fall, a number of perfectly pleasant bushes reveal an alter ego that is surprising and far more fabulous than might be expected.

With bright colors and energetic form, stems are a winter garden essential. Adding a distinctive intensity to the scene, these are plants that don't just stand up to wind and cold weather, but positively relish it; bending in the stiffest breeze, unconcerned in heavy rain and set off to perfection by frost and snow.

When it comes to choosing varieties for the garden, there is a range of options depending on location and conditions, but when it comes to sheer brilliance it is the shrubby species of *Cornus* that lead the parade. Red *Cornus alba* 'Sibirica' loves damp soil, *C. sanguinea* 'Anny's Winter Orange' will suit a well-drained spot, and vigorous *C. sericea* 'Flaviramea,' with its zingy, acid green twigs, is fairly unfussy.

Ornamental brambles include sophisticated white-purple *Rubus biflorus* and chalk-white ghost bramble, *R. thibetanus*, while the arching stems of Japanese wineberries, *R. phoenicolasius*, are covered with a fuzz of fine orange prickles that capture both light and frost delightfully.

Willows are similarly colorful, and while some species do make substantial trees, others are smaller,

Clockwise from top left *Rubus biflorus* is an elegant, if prickly, addition to the garden. The copper stems of *Rubus phoenicolasius* also have a rusty halo of thorns. The showy yellow canes of *Phyllostachys aureosulcata* f. *spectabilis*. Some willows such as *Salix alba* var. *vitellina* can be very striking. The bold, black rods of bamboo *Phyllostachys nigra*. *Cornus sanguinea* 'Anny's Winter Orange' is outstanding for color.

Making the most of colored stems

- Site shrubs or trees where they will catch the low sun, to set the colors on fire.
- Plant *Cornus* and *Rubus* species in blocks or layers to maximize impact.
- Underplant with contrasting bulbs or alongside companions that enhance the stem color.
- Clear the stems for a vertical statement by trimming back bamboo leaves, thinning stems, or raising the crown of trees (see p.69).
- Cut the twigs and bring them indoors to use in wreaths and arrangements.

such as red-gold *Salix alba* var. *vitellina* 'Britzensis' and *S. × sepulcralis* 'Erythroflexuosa,' which has twisted stems. In many cases, size can be controlled or moderated by annual or biannual pollarding—which has the advantage of allowing an interestingly gnarled trunk to develop—while cultivars of *Salix alba* var. *vitellina* are often grown as a shrub in smaller spaces, and coppiced regularly.

Although its segmented canes are handsome, bamboo is a bit of a brute, and it should be chosen carefully as some species can run out of control. Less invasive varieties include purple-black *Phyllostachys nigra* and attractive, upright *Fargesia* Red Panda, both of which form clumps, but it is still worth restricting sideways growth by planting in a large container, or using some other impenetrable barrier.

Acers are extraordinarily good value in the garden at all times of year. In addition to three seasons of foliage color, they have interestingly textured stems in various shades of green, yellow, silver, and cinnamon (see pp.68–69). The younger shoots of *Acer palmatum* 'Sango Kaku' are coral red beneath citrus fall leaves, while those of *A. conspicuum* 'Phoenix' are a brilliant orange-pink. These acers will tolerate light pollarding to restrict their size and encourage the colorful stems, or they can be allowed to develop naturally.

Getting the best color

When plants are grown for colored stems, the key to success is usually in pruning (see p.130). Young wood has the best color, and cutting shrubs back regularly will stimulate plenty of bright new growth, while also helping to keep more vigorous varieties under control.

Willow: grow as a coppiced shrub or pollard as a tree, cutting back hard annually or every two or three years.

Cornus: cut back the oldest stems in rotation, taking out one stem in three in spring at the base. With cultivars of *Cornus alba*, once the plant is well established, winter impact can be maximized by cutting the whole plant down almost to ground level, then feeding and mulching to encourage new, straight growth.

Rubus: ornamental blackberries are vigorous and thorny, so pruning requires thick gloves. You can either take out about a third of the stems each year—suitable for *R. biflorus*, and *R. phoenicolasius* (Japanese wineberry), which produces fruit on two-year-old canes—or, if a plant requires controlling, cut it down to the ground annually. This works well for *R. cockburnianus.*

Other plants, shrubs, and trees: bamboo can have surplus stems removed as required; use these as canes elsewhere in the garden. In the case of trees other than willow, pruning should be a more considered undertaking. Acers and birches need a careful approach so that their attractive shape is maintained, but *Tilia cordata* 'Winter Orange' can also be pollarded, so it pays to get to know the plant in question and proceed carefully to see what the plant will and will not tolerate.

Right The bold new growth on this row of pollarded willow trees is particularly effective against the calm, winter-gray woodland.

Evergreen Climbers, Trees, and Shrubs

Plants that keep their leaves in winter remain handsome, structural, and interesting. Many gardens, therefore, have an evergreen base layer, made up of things that are cheap, have a long season of interest, and are hard to kill, but when adding to a planting scheme, picking sturdy varieties that are also exciting will improve the outlook enormously.

Trees and large shrubs

With its glossy leaves and scarlet berries, common holly (*Ilex aquifolium*) is peerless, but alternatives include *I. aquifolium* 'Bacciflava,' which has bright yellow berries, or one of the variegated forms such as *I. × altaclerensis* 'Golden King,' yellow-berried *I. aquifolium* 'Argentea Marginata,' or 'Silver Milkmaid.' Holly is dioecious and these female plants bear berries, but male plants, such as *I. aquifolium* 'Silver Queen,' do not.

Other useful small trees include the strawberry tree, *Arbutus unedo*, which has cinnamon-colored bark, red fruit, and cream flowers, while white-bloomed *Eucryphia* thrives in full sun, as long as its roots are moist. Bay (*Laurus nobilis*) clips well, the bark of *Eucalyptus* species flakes attractively, and *Parrotia persica* also has good fall color.

In a modern or tropical scheme, *Fatsia japonica* excels: it may look exotic but it is very easy to grow. Hardy palms include the Chilean wine palm (*Jubaea chilensis*) and *Trachycarpus fortunei*, both of which look esoteric in snow.

Good shrubs, meanwhile, are legion. Fragrant *Mahonia* and *Sarcococca* are stalwarts, and in a shady spot, moisture-loving camellias can flower from late fall to early spring, depending on variety.

Reliable plants are often belittled, but for those who reject *Skimmia*, *Photinia*, and *Pieris*, more exotic subjects exist, such as *Gaultheria* species, which prefer moist, acid soil, together with *Euphorbia mellifera*, *Callistemon rigidus*, and *Azara microphylla*, which has vanilla-scented yellow flowers in late winter.

Climbers

Ivy's reputation as a thug precedes it, but although wild forms might be rampant, there are a number of cultivars with finely divided leaves, subtle variegation, or a compact habit. It also lends itself to trailing or cascading, as well as climbing. Like most plants, ivy will grow in a more restrained manner in suboptimum conditions, so confining it to a container will help keep it controlled. This is ideal for smaller areas such as courtyards, balconies, and roof gardens.

Clematis, too, has a number of useful evergreen varieties for the winter garden. In a large space, *C. armandii* will stretch out long stems with big, leathery leaves, while in a sheltered spot, cultivars of *C. cirrhosa* are a good choice, flowering in winter and early spring, alongside *C. urophylla* 'Winter Beauty.' *Clematis × cartmanii* 'Avalanche,' meanwhile, is compact enough for a small garden and while it doesn't flower until late spring, it is less vulnerable to slugs than deciduous forms are.

A number of other evergreen climbers have a place in the winter garden, even if their flowers are absent. *Solanum jasminoides* and *Trachelospermum jasminoides* can scramble up a trellis, or pergola, as can rampant *Holboellia* species. There are also many shrubs that can be trained against a protective wall to great effect, such as *Garrya elliptica*, with its long silver catkins.

Clockwise from top left Variegated holly, *Ilex aquifolium* 'Silver Milkmaid.' Ivy is a good plant for wildlife; *Clematis urophylla* 'Winter Beauty' has bell-shaped blooms; *Garrya elliptica* looks lovely trained against a wall; unusual yellow-berried *Ilex aquifolium* 'Bacciflava'; *Hedera* is a very diverse genus that has strong winter foliage.

Pick of the Best Evergreens

While many plants have hidden strengths, evergreen plants can make a winter garden, so it is worth putting some serious thought into what will improve an existing arrangement, or strengthen a new one.

When choosing evergreens (and plants generally), it pays to look around to see what is available. Frequently used plants should not be discounted: they are likely to be widely available precisely because they are good and reliable. For those that are already growing in the garden, a bit of a trim and some underplanting may revitalize them completely.

If the selection in the garden center seems restrictive, it is worth recognizing that for every dull plant, there are dozens more that are interesting, exciting, and just as easy to grow if you know where to look. Visit small growers and search online; check out botanic gardens, national collections, and arboreta. Talk to the specialists and the enthusiasts and do some research. Adding even a few things that are different and interesting will elevate the planting scheme, and your garden will be all the better for it.

Unusual varieties and subtle solutions

For the adventurous of spirit, all sorts of plants can be used as an antidote to the run-of-the-mill winter garden palette, particularly in smaller spaces and shady spots. Elements of Japanese style can be introduced with hummocks of moss or mossy rocks and small evergreen ferns such as *Asplenium trichomanes*, *Blechnum spicant* or, in a sheltered spot, the maidenhair fern, *Adiantum venustum*.

There are also dwarf varieties of ivy, such as *Hedera helix* 'Minima' or *H. helix* 'Dyinnii,' which is useful for alpine troughs and can be grown alongside either black or green forms of *Ophiopogon*. Remarkably hardy is *Euphorbia myrsinites*, which

has four exciting seasons of interest and can be teamed with small bulbs and succulents such as sempervivums. Grown in a container in a sheltered spot, the gin and tonic plant, *Eriostemon myoporoides*, flowers for a long period over winter and into spring.

It pays to assess the available space and then be inventive. If something larger is required, there are many shrubs and small trees that are less frequently used, such as *Pseudopanax ferox*, *Chamaerops humilis*, *Eriobotrya japonica*, and *Chamaecyparis pisifera* 'Golden Mop.' A trip round an arboretum or other specialist collection will provide food for thought.

Box and its alternatives

Box is a classic garden evergreen, and whether topiarized, clipped into a hedge, or just growing free, it is the first choice for year-round structure for most people. Yet with the arrival of box tree moth (*Cydalima perspectalis*) and the fungal menace that is box blight, now pretty much endemic, it is no longer the healthiest or best-looking hedging plant around.

Considerable research has been done on alternatives to *Buxus*, particularly a long-term trial at the RHS garden at Wisley in Surrey. The aim has been to find a range of other plants with small leaves and a compact habit, which responds well to clipping. Ideally, these will perform in a similar manner to box and can be used to create gardens in the traditional style.

So far, *Taxus baccata* 'Repandens' and *Berberis darwinii* 'Compacta' are showing promise and *Ligustrum delavayanum* is an effective substitute, while *Ilex crenata* and *Osmanthus* species are also worth a try. However, with the changing climate, drought, heat, and periods of severe cold are also having an impact. As a result, it is worth considering your current local conditions, and the kinds of weather events that are likely to become more frequent over the next few years, and bear that in mind when planting something new.

Right With plenty of diverse and interesting evergreen foliage in a garden, it would be hard to pinpoint this as a winter scene, were it not for a few bare-stemmed, tell-tale birches.

Designing *with* Evergreens

Clever use of evergreen foliage can keep a garden looking good all year round. Contrasting different plants within a defined framework, and employing seasonal highlights will result in a design which is both bold and harmonious.

In smaller gardens, a pared-back palette of hardworking plants is often the best way to create a space that works effectively, and this is where evergreens excel. Repeating topiary, small conifers, heathers, or big-leaved bergenias can create layers of movement and almost infinite detail, managing to be interesting while still giving an impression of restraint.

Combinations can be fairly loose, or classical influences might result in formal arrangement of statuary and foliage. Taking inspiration from the simplicity of Japanese garden style, meanwhile, can be extremely effective. In Japan, gardening is an art form that takes its cues from the natural world to create a landscape that is both beautiful and rich with meaning. Key elements include water, or the idea of water as a life force; stone, to represent features such as mountains; and plants—many of which are evergreen.

It is a style that harnesses the things that are abundant in winter and allows creative exploration of ideas such as stillness, enclosure, and balance, while remaining very flexible on a practical level. Rocks and raked gravel are structured; a scene made up of shades of green is restful to the eye, and limiting color to different cultivars of a single plant, such as hamamelis, snowdrops, or hellebores adds intrigue and considerable depth.

Right The restrained palette and allegorical landscape of a Japanese garden is usually less brash than the British evergreen gardening tradition.

Lawns and Grasses

In recent years, the role of grass in a garden has changed. It is no longer seen just as a neat suburban lawn or a verdant apron between a stately home and its rolling acres. New fashions have emerged: gardens have got smaller, and grasses in their widest sense have been embraced as distinguished and adaptable garden plants, in a genre of their own.

The humble lawn has expanded its remit, tweaked to accommodate wildlife and plants formerly known as weeds in an aspirant step toward meadow status. While stripes rolled on freshly cut grass still have some champions, it is a style that seems increasingly anachronistic within a spectrum of lawns that are formal, informal, ornamental, herbal, or even, regrettably, artificial. Meadows have gained popularity in gardens of all sizes, and prairie planting goes a step further, bringing together grasses and perennials that are both low maintenance and long-lasting.

Grasses are supremely versatile. In both design and function they can be just as significant in a winter garden as they are in summer, and for that alone, they are worthy of attention.

Grasses *in* the Garden

Functional, ornamental, and elegant, grass is a fundamental garden element that can be used in a plethora of ways. As a backdrop to family living or a chic planting scheme it excels, yet it is the wilder, more playful incarnation of meadow or prairie that has more recently updated its appeal.

The conventional lawn is often a significant part of a garden, but its nature depends on the philosophy and lifestyle of the household in question. It may be a pristine sward, the gardener's pride and joy, tenderly nurtured until it is the epitome of green and vibrant perfection. Alternatively, it may be a more haphazard and utilitarian affair; just a patch of grass, subject to the rough and tumble of everyday life and flattened underfoot on the way to the washing line or compost bin.

In either case, grass can take up a considerable amount of our outdoor space, and winter brings its challenges. Frequent trampling will cause the ground to become compacted, reducing drainage and leading to waterlogging, while walking on grass in wet or frozen conditions can rapidly cause muddy puddles and bald spots—all of which raises the question of whether a lawn is desirable, or even necessary.

Yet, considered as part of the overall garden design, the elegant simplicity of grass is a huge asset.

Left Walking on the frozen lawn is best avoided as it will cause the leaves to snap and the cells within them to rupture, leaving a trail of damage behind after the thaw.

A lawn represents an empty space and it is somewhere that the eye can pause. It allows focal points, sculpture, or a breathtaking view to command one's full attention, while areas of busy, energetic planting are offset by its stillness.

On bright, cold days, a simple lawn is transformed, becoming a foil for shadows and textures. It is like a piece of art that gently evolves, as frost melts and reforms, and as the light is momentarily softened by clouds rolling by. When the sun is at its zenith, the lawn may be imprinted with a crisp tracery of branch shadows, black on green; but as it sinks toward the horizon, trees are rendered in smudgy, impressionistic, charcoal lines against a soft haze of purple and gold.

Areas of cut grass can be a canvas for botanical expression, too. Small bulbs are ephemeral and, when planted in clusters under trees or in a naturalistic swathe, they create a delicious splash of color (see p.140 for planting information). This optimizes use of space and boosts winter interest, and by the time a regular mowing regimen is reinstated and outdoor activities resume, the flowers will have died back and vanished almost without trace.

Caring *for* Lawns *in* Winter

While the grass is usually the focus of attention, it is also important to take care of the soil, and especially so in winter. Overuse will cause compaction and muddiness, while too much mowing can do more harm than good. Above all, the lawn needs a rest.

General care

Avoid compaction: normal, healthy soil contains air spaces, which allow roots to breathe and rainwater to be absorbed like a sponge. Regular or heavy trampling, particularly when the lawn is wet, squeezes the soil particles together and pushes the air out. As a result, plants struggle and water either collects on the surface or runs off into adjacent areas, neither of which is desirable.

Let the light through: rake off leaves and twiggy debris to avoid the grass underneath being smothered and killed. This material will break down in a compost bin or it can be bagged up until it turns into leaf mold, but the easiest option is to deposit it in an unobtrusive corner, behind some shrubs, where the wildlife can take advantage of it while it rots away.

Mow less frequently (or not at all): grass grows more slowly in winter, so as long as the weather is fairly cool, mowing can be reduced to an occasional trim if absolutely necessary. Allowing grass to remain a little longer will trap rainwater, giving it longer to soak into the ground; this reduces localized flooding and helps to replenish the soil so it can better tolerate dry periods later in the year.

Stop feeding: feeding is unnecessary in winter, so wait until the weather warms and the grass is actively growing. Applying fertilizer now is not just a waste of time and money; in heavy rain the nutrients are likely to run off in surface water or leach out of the soil. This risks contaminating watercourses and causing algal blooms, in a process known as eutrophication.

Restoration and maintenance

By the end of the summer, lawns can be looking a bit worse for wear, with mower-scrapes, bald patches, compaction, and drought. Generally, it is best to tackle renovation when the climate is forgiving, in mid-to-late fall or midspring, to give the grass the best chance of recovery.

Three-step plan for recovery

- Aerate your lawn to allow the roots to breathe. You can do this with a hollow-tine aerator, or, if the area is small, plunge a narrow-tined garden fork up to 4in (10cm) deep into the soil to create air spaces.
- Scarify the lawn to remove excess thatch. This is the layer of dead grass and other material around the roots; it is not usually a problem when most of the grass species are slow-growing but when grass is fast-growing or overfed, it can accumulate. Rake the surface of the lawn, as if combing its hair, to remove the surplus stems and let in light and air. The amount of effort this takes will depend heavily on the thickness and health of the grass.
- Reseed bald patches, picking a time when the weather is moist but not too cold. Always choose the right seed mix for the conditions in your garden; there are seed mixes for all situations and soils, from sun to shade and general wear to heavy use. Slow-growing and low-maintenance formulations that include nongrass species are also available.

Top right Raking fallen leaves off the lawn in fall and early winter helps to keep the grass healthy over winter.

Bottom right Longer grass is more resilient, so lay off the mowing in winter and give the lawn a rest.

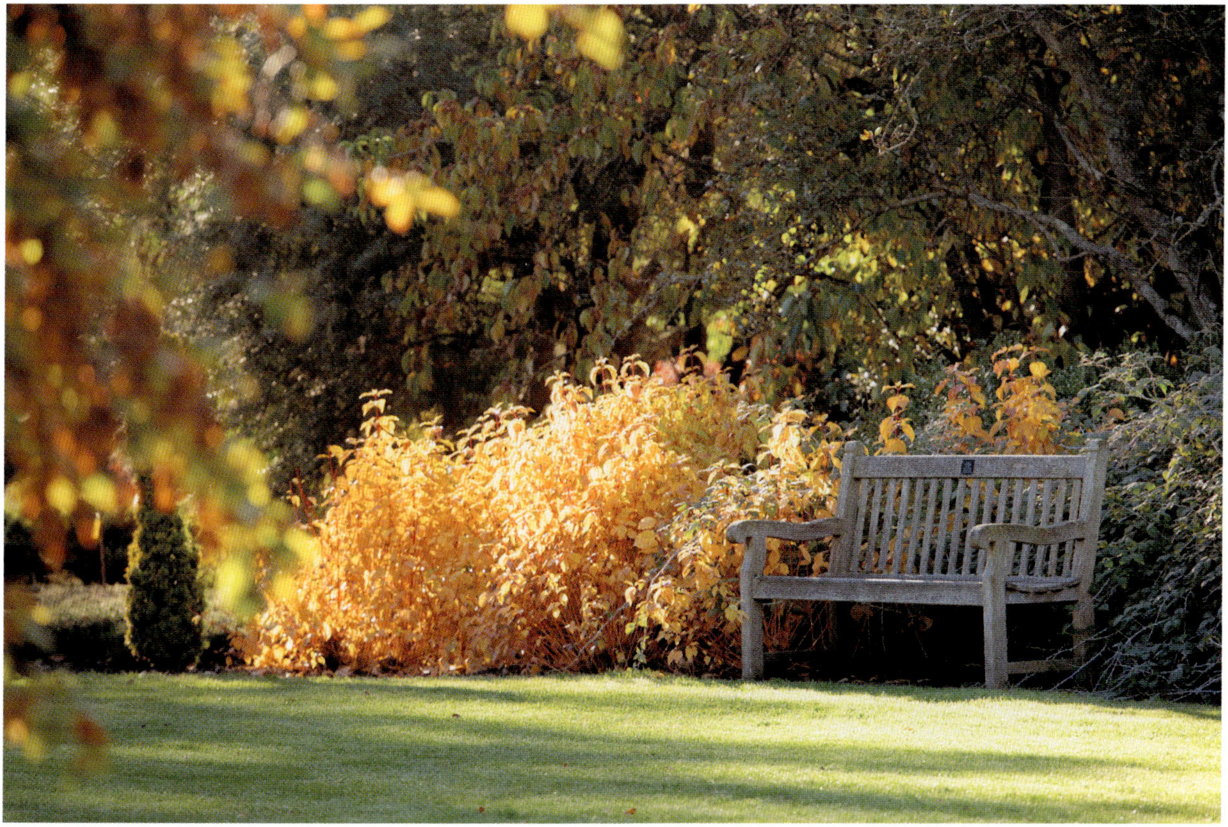

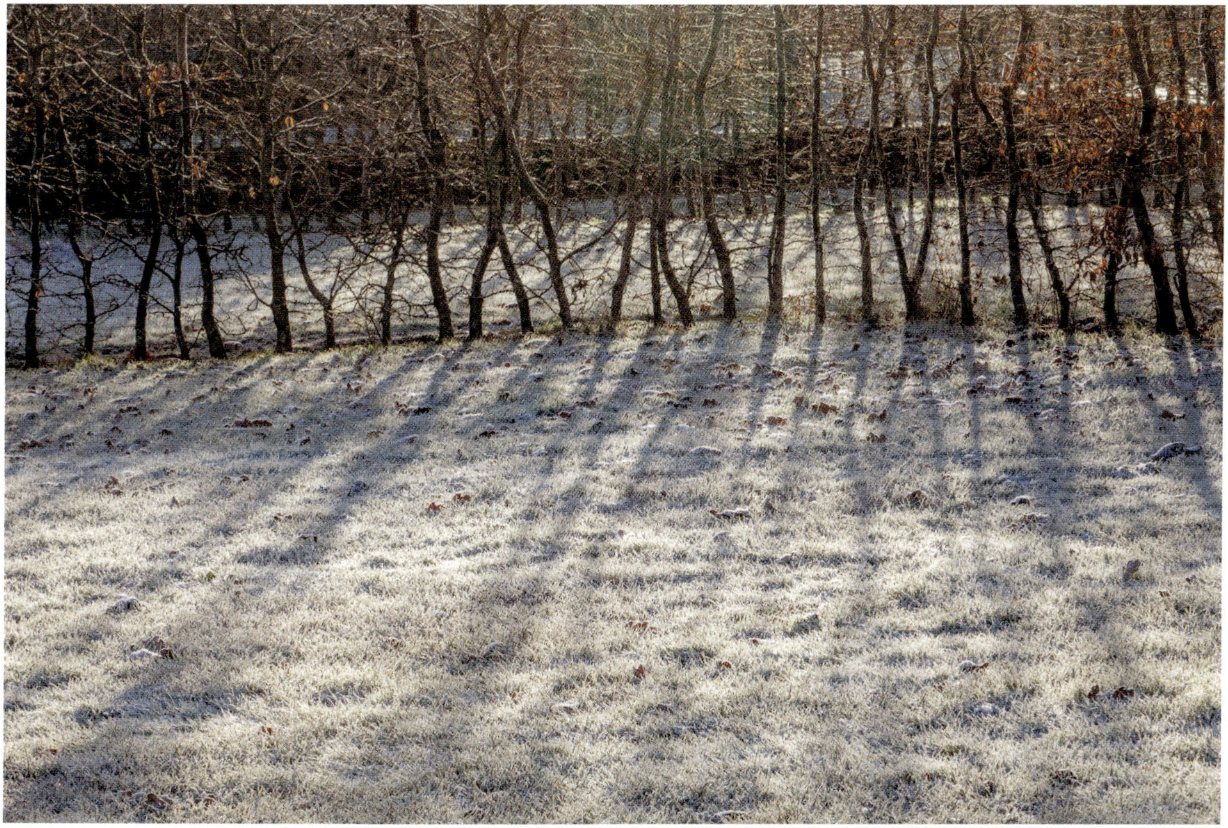

Meadows

Bringing a meadow into the garden is gaining popularity, and for good reason. Grass left shaggy and infiltrated with a variety of other plants might lack order and neatness, but it makes up for that in terms of interest over a long period, becoming something that is neither a clipped lawn, nor a riotous modern prairie, but occupies a large and rather nebulous space in between.

Rougher and more glorious than a conventional lawn, meadows are a forgiving feature to include in a design. Although the term was traditionally used to describe open, flower-rich grassland, it has more recently been adopted by gardeners to cover a spectrum of planting options. This can range from welcoming wild flowers into a lightly managed patch of grass in a small urban plot, to something that is more consistent with a traditional managed landscape, such as a hay meadow populated with native species. There are also horticultural variants, which involve putting together a mix of ornamental plants and grasses to create particular effects.

In all these cases, the increased range of plants and a lighter management regime help to support wildlife. Compared to grass alone, a mixed meadow planting will provide a long period of pollen and nectar for insects, seeds to feed birds and small animals, and all sorts of nesting material and hibernating places as well. There are also other environmental benefits, such as improved water management, as longer vegetation will help trap rainfall and slow the rate of surface runoff (see p.44).

The core idea is that a garden meadow should be managed in the same way as a traditional hay meadow. This means allowing it to grow until the flowers have faded and the seeds have dropped in mid- to late summer, then mowing it and raking off the plant material, which can then be composted.

Mow the lawn a few more times before early spring, raking off the grass on each occasion. This replicates the season of grazing that follows haymaking. Removing the organic matter in this way reduces the level of nutrients in the soil, which gradually tips the balance in favor of wild flowers rather than promoting lush grass. For the same reason, it is not usual to feed an ornamental meadow, as most soils are rich enough already.

The meadow in the garden

Meadows are a natural patchwork, which makes them highly versatile. The varied community of plants means that they are more resilient to challenges such as drought, and they can be adapted to suit various situations and move with the seasons.

For aesthetic purposes and to save time, paths or clear zones can be mown through longer grass as required, with a clear-cut done a few weeks or months later. Thus, frosted seedheads can be enjoyed while they last, while still making space for spring bulbs, followed by camassias and cow parsley in early summer.

Specimen trees add height and sculptural form and can be underplanted with swathes of narcissi for spring, or colchicums for fall. Long grass could also be combined with well-spaced fruit trees, to create an ornamental orchard, or echo in microcosm the historic European custom of Streuobstwiesen. This is a form of wood pasture where fruit trees are grown in meadowland, enabling a crop of apples or cherries, grazing for animals, a harvest of hay and sometimes honey too, in an ultraefficient and sustainable way.

Right At Le Jardin Plume in Normandy, France, tightly mown paths intersect regular squares of longer grass, which are offset by trees and feathery ornamental planting. When these meadow areas are cut late in the season, their imprint and quiet formality remain, but there is also space for defined swathes of crocuses, early daffodils, or *Fritillaria meleagris* to put on a show, before the grass starts to grow once more.

Ornamental Grasses

Adaptable and long-lasting, ornamental grasses can play a number of roles in the garden. Teamed with perennials or in massed, single-variety plantings, grasses become fuller during summer, to provide useful late interest alongside dahlias, Michaelmas daisies, and salvias. The ones that are evergreen then carry gaily on into winter, but even those that are technically deciduous can often remain spectacular in their skeletal form.

Ornamental grasses have a charming habit of getting better and better as the year wears on—and the taller they are, the more marked is the improvement. From hummocks of bristling green in spring, the blades extend to assume their true, expansive character, replacing and refreshing tired old growth. In late summer they typically hit a peak when the flowers arrive, only to reinvent themselves again during fall, for an extended stay of interest.

Some grasses and sedges such as *Anemanthele lessoniana* (pheasant's tail grass) and *Carex* cultivars are dramatic in form, while remaining relatively modest in size, yet there are others whose mighty, vertical stems are a striking presence in the border. Though their foliage may fade and wither, the airy plumes of miscanthus and pampas grass stand tall and airy, catching light and frost while providing a sense of texture and fluidity, in among trees and evergreens.

Left Imposing even in winter, ornamental grasses are improved still further by a hard frost.

Overleaf With shimmering plumes, sparkling cascades, and silver fountains, grasses bring energy and movement to the chilly garden in a range of different ways.

Of all the plants in the winter garden, grasses are perhaps the closest in form to fireworks. The colors may be calming arrays of taupe and bronze, soft lime and green, but the shapes are energetic. There are fountains of *Pennisetum alopecuroides*; shooting, vertical flames of *Calamagrostis* and *Deschampsia cespitosa* like showers of rain; *Pennisetum macrourum* is like exploding sparks and, backlit by low sun, the seedheads of *Miscanthus nepalensis* are reminiscent of screaming spiraling sparks, frozen in time.

Grasses for smaller spaces

In smaller spaces and formal gardens the largest grasses may be less suitable, but there are plenty that are more compact. These lend themselves to a design in various ways; they work well in a border or as part of a mixed container, and they can also be used as a unifying theme or as checkerboard-effect groundcover.

Taller cultivars, such as *Panicum virgatum* 'North Wind,' may still be used sparingly, perhaps to create a screen or introduce an airy vertical accent at the back of the border, while a grass such as *Stipa tenuissima* with its neat clumps and gauzy tassels may be moderate in size but loses nothing in style. Repeated plantings of smaller, evergreen grasses such as punky *Carex oshimensis* 'Evergold' can be used to bounce the eye around the space and fill bare borders.

Underplanting deciduous shrubs or colorful stems with tussocky *Carex comans* bronze-leaved or *Anemanthele lessoniana* creates the soft, fluid effect of a bleached winter meadow, and these also last well alongside the dried pods and seedheads of an herbaceous planting scheme. Green *Carex morrowii*, meanwhile, provides a quiet, low foil for topiary—not unlike a traditional lawn.

In winter containers, grasses can be grown by themselves as a statement plant, or used as a softer backbone to seasonal color and bedding. For the specialist with an eye for aesthetics, well-behaved varieties in an alpine trough make a good foil for evergreen specimens and small, choice bulbs, too, providing long-lasting interest and filling the space before and after the main event.

Prairie Planting

There is something hugely evocative about the prairie planting style popularized over the last few decades by designer Piet Oudolf. The repeating swathes of grasses and flowering perennials are sculptural and diaphanous, creating an effect that is gentle yet purposeful and measured in its rhythm, playing on the psyche like a familiar piece of music.

Referencing the wild grasslands of the American Midwest, prairie planting is characterized by loose drifts of naturalistic perennials and grasses that combine to form an airy, dreamy landscape with a wild aesthetic. While this works particularly well in large spaces, there are elements that can be adapted to more modest domestic situations, too, in a style that is sometimes known as the new perennial movement.

This is a highly adaptable concept. Although based on the flora and sunny, open environment of a North American plain, it can include plants from all over the Americas, together with species that like the same conditions, particularly those from the Mediterranean and South Africa.

From moor to garrigue, these natural landscapes are often composed of a characteristic local flora, with individual plants repeated by the hundred thousand. Though each may be a small element, the result is striking, and prairie planting draws on this, combining plants in swathes or on a matrix system to build up a layered effect that swells and recedes with the turning season.

In the garden, it is relatively straightforward to develop the theme in a sunny border, or even to replace a conventional lawn with something more

Right Choosing plants with strong, upright stems and durable seedheads allows a prairie planting scheme to have garden impact well into winter.

dramatic. The palette of available plants is extensive and impactful, and as a scheme it is really no more high maintenance than any other grassed area or perennial border. In addition, it is supremely wildlife-friendly and a rich source of the grasses and seedheads that lend themselves to winter floristry (see p.181).

Prairie garden plants are often resilient, and those that also have a strong shape will excel. Try grasses such as *Calamagrostis brachytricha* 'Mona' and *C. emodensis*; compact *Cortaderia selloana* 'Pumila,' *Miscanthus sinensis*, and *Panicum virgatum* cultivars such as 'North Wind' or 'Cheyenne Sky,' while *Bothriochloa bladhii* also has striking fall color. The winter cones of *Rudbeckia* and *Echinacea* cultivars are both attractive and persistent, while points and punctuation can be succinctly added by the rounded seedheads of *Sanguisorba*, *Phlomis*, and *Cephalaria*; these contrast well with the flat forms of *Achillea* and *Hylotelephium spectabile*.

The prairie garden in winter

In late summer, a new perennial garden is a magnificently billowing vision and a hive of pollinator activity. But in winter it shows another part of its character; as the flowers dwindle, stems and seedheads stand tall in a landscape of elegantly wasted skeletons. As colors fade to muted tones of taupe, sand, biscuit, and coffee, the low sunlight picks out blond and silver details. Shade and frost pare the design back further to a virtual monochrome, while fog lends an ethereal softness.

Planting a prairie garden

The key to this style of design is choosing a restricted palette of plants, and then planting multiples of each variety in swathes or blocks to create a flowing effect in which the unevenness of odd numbers looks more natural. Always use enough plants to fill the space. If the budget is tight and time is not an issue, it is perfectly possible to grow a prairie border from seed.

First of all, assess your site. Prairies are naturally sunny places with free-draining soil, and many traditional prairie plants cannot grow in any other conditions. In a garden blessed with some dappled shade or damp ground, adjust the scheme accordingly by choosing species that will do well in the conditions available.

The next thing to do is choose the best plants in your scheme. There should be plenty of grasses but it can also include a range of other perennial plants. For an effect that will last long into winter, choose strong stems and robust seedheads; good performers include cultivars of *Miscanthus sinensis*, *Calamagrostis*, *Eupatorium*, yarrows, and coneflowers.

Well-prepared soil is essential for the plants to put on a strong display, so weed it and loosen it as necessary. Prairie plants like good drainage, so on heavy soil make this a priority by adding plenty of organic matter to improve the conditions.

When it comes to laying out the plants, arrange your chosen species in blocks or drifts, so each variety is massed together for maximum impact. Aim for repeated colors, shapes, and species, bearing in mind both height and the degree of solidity or translucence that each plant brings, and how this will affect the final impression.

Once the planting is finished, water well and continue to do so while the plants establish. Toward the end of the winter season, when the remaining stems become battered beyond redemption, cut the whole lot down to ground level, pile up the trimmings where the wildlife can take advantage of them, and wait for the show to repeat next year.

Clockwise from top left *Calamagrostis brachytricha* lasts well into winter. *Monarda* seedheads catch and hold the frost. *Sanguisorba* 'Cangshan Cranberry' adds a pretty punctuation. Compact *Miscanthus sinensis* 'Gnome'. *Phlomis tuberosa* 'Amazone' contributes a bold, structural element. *Achillea filipendulina* 'Gold Plate' is subtle in senescence.

Glossary

Bract A modified leaf structure, often below the reproductive parts of the plant. Bracts can differ from the rest of the foliage and they may be striking or brightly colored.

Coppice, coppicing The practice of cutting a shrub or tree back to ground level, in order to stimulate fresh new growth for commercial or aesthetic purposes.

Dark sky principles A set of principles for responsible use of outdoor lighting to reduce light pollution and save energy. See the International Dark-Sky Association for more information.

Detritivores Organisms that feed on dead organic matter, contributing to decomposition and the cycling of nutrients.

Galanthophile A lover of snowdrops.

Garrigue Open, shrubby vegetation typical of the Mediterranean; a type of scrub made up or aromatic and thorny plants, found on well-drained calcareous hillsides, near the sea.

Graft union The join between the rootstock and the top-wood that conserves the desired cultivar or characteristic in a grafted plant.

Gymnosperms A group of plants that produces naked or unenclosed seeds. This includes cycads, Ginkgo, and conifers.

Eutrophication The process by which a body of water such as a river or lake becomes enriched with excessive nutrients, typically nitrogen and phosphorus from fertilizers. This causes uncontrolled growth of algae and harms aquatic ecosystems.

Etiolate Where a plant grows tall, thin, and pale through lack of light.

Pollard To cut off the branches of a tree to promote compact and bushy growth, and to help manage its size and spread.

Potash The horticultural term for the element potassium (K) in water-soluble form. The name comes from the original practice of collecting wood ashes in a container.

Rule of thirds Where a composition such as an image, structure or design is divided evenly into thirds, both horizontally and vertically.

Sepals The individual leaflike structures that enclose the flower in bud, and collectively make up the calyx.

Thermal mass The property of a material to absorb, store, and release heat. The high thermal mass of concrete, brick, and water means that they provide a buffer against rapid changes in temperature.

Xerophyte A plant which is adapted to arid conditions.

Index

Picture Credits

The publisher would like to thank the following sources for their kind permission to reproduce their photographs:

Key: t = top c = center b = bottom l = left r = right

GAP = GAP Photos
MMGI = Marianne Majerus Garden Images

Page 2 GAP: Howard Rice, Cambridge Botanic Gardens; **4-5 © Andrea Jones Photography**, courtesy Culzean Castle (NTS); **6tl and bl Jason Ingram**: RHS, **tr GAP**: Adrian Bloom, **br GAP**: Martin Staffler; **10-11 GAP**: John Glover; **12 David C Phillips**; **15t Jonathan Buckley**: design Alan Titchmarsh, Hampshire; **b MMGI**: Marianne Majerus: Gravetye Manor; **17 Clive Nichols**: Pettifers Garden, Oxfordshire; **18-19 MMGI**: Marianne Majerus: St Timothee, Berkshire; **21 Ned Lomax**; **22-23 GAP**: Jerry Harpur; **24tl GAP**: Andrea Jones, **tr Jonathan Buckley**; **b Clive Nichols**; **26-27 OTTO**: Stephen Kent Johnson, Landscape Design: LaGuardia Design Group, Architecture: Architecture Plus Information (A+I), Interiors: Unionworks; **28-29 GAP**: Annette Lepple, garden: Mas de Bety; **30 Jonathan Buckley**: design John Massey, Ashwood Nurseries; **33t GAP**: Howard Rice, **b GAP**: Elke Borkowski; **35tl GAP**, **tr GAP**: Carole Drake, Rustling End, Julie and Tim Wise, **b GAP**: Friedrich Strauss; **36-37 Rob Cardillo**; **38l and r Ray Cox Photography**; **41tl Shutterstock**: Paul Roedding, **tr GAP**: Jenny Lilly, **b Alamy**: Erik Agar; **42-43 MMGI**: Marianne Majerus, St Timothee, Berkshire; **45tl Rachel Warne**, Beth Chatto, Water Garden, **tr Annaick Guitteny**, Domaine du Ooievaar, **b Clive Nichols**; **46-47 Ray Cox Photography**; **48-49 Jason Ingram**; **50 GAP**: Richard Bloom; **53t and b GAP**: Richard Bloom, designer: Tom Stuart Smith, Cogshall Grange; **54tl GAP**: Elke Borkowski, **tr and b GAP**: Joe Wainwright, Wollerton Old Hall, John and Lesley Jenkins; **56-57 Clive Nichols**, Wollerton Old Hall, Shropshire; **59tl GAP**: Richard Bloom; **tr GAP**: Fiona Rice, **b Jonathan Buckley**, Wol Staines, Glen Chantry, Essex; **60-61 GAP**: Carole Drake, Simon Dorrell, Ashley Farm, Roger and Jackie Pietroni; **62 GAP**: Richard Bloom; **64-65 GAP**: Annie Green-Armytage; **67tl Clive Nichols**, **tr GAP**: Richard Bloom, **bl GAP**: Roger Mabic, **br GAP**: Heather Edwards; **68tl GAP**: Carole Drake, Cleave Hill, Membury, Andy and Penny Pritchard, **tr GAP**: Karen Chapman, Le Jardinet, **b GAP**: Tim Gainey; **72-73 Clive Nichols**, Ellicar Gardens, Nottinghamshire; **74 GAP**: Richard Bloom, The Winter Garden, Bressingham Garden, Norfolk, design Adrian Bloom; **77t and b Jason Ingram**; **78tl MMGI**: Marianne Majerus, Matt Keightley, **tr MMGI**: Marianne Majerus, design Julie Toll, **b GAP**: Rob Whitworth, Knoll Gardens; **81tl and bl Jason Ingram**: RHS, **tr GAP**: Adrian Bloom, **br GAP**: Martin Staffler; **82-83 MMGI**: Marianne Majerus, Markshall Estate, Essex; **85t and b MMGI**: Marianne Majerus, design Annie Pearce;

86-87 Jonathan Buckley; **88 Mette Krull**; **91tl and b Clive Nichols**; **tr MMGI**: Marianne Majerus, design Julie Toll; **92 Jason Ingram**; **94 GAP**: Friedrich Strauss; **97tl GAP**: Pernilla Hed, **tr Clive Nichols**; **bl Jason Ingram**, **br Clive Nichols**, design Kathy Brown; **99tl Jason Ingram**, **tr MMGI**: Marianne Majerus, **b GAP**: Friedrich Strauss; **100-101 GAP**: Joe Wainwright, Wollerton Old Hall, designers: John and Lesley Jenkins; **102 GAP**: Richard Bloom; **105 GAP**: Lynne Keddie; **107tl GAP**: Howard Rice, **tr Emily Daw**, Lower House Garden, **bl GAP**: Howard Rice, **br GAP**: Carole Drake, Cleave Hill, Membury, Andy and Penny Pritchard; **108t GAP**: Joe Wainwright, Wollerton Old Hall, Shropshire, John and Lesley Jenkins, **b GAP**: Matteo Carassale, designer Antonio Perazzi; **111 GAP**: Brent Wilson, **b Richard Bloom**, design Sue Townsend; **112-113 MMGI**: Bennet Smith, design: Julia Jarman; **115tl Clive Nichols**, Wollerton Old Hall, Shropshire, **tr Jason Ingram**, **cl Rob Cardillo**, **cr GAP**: Howard Rice, Madingley Hall, Cambridge, **bl MMGI**: Bennet Smith, design Charlotte Molesworth, **br Jason Ingram**; **116-117 GAP**: Richard Bloom, The Winter Garden, The Bressingham Gardens, Norfolk, Adrian Bloom; **118 MMGI**: Marianne Majerus, Anglesey Abbey, Cambridgeshire; **120-121 Emily Daw**, Lower House Garden; **122tl Clive Nichols**, **tr GAP**: Howard Rice, Cambridge Botanic Gardens, **cl Jonathan Buckley**, **cr Jonathan Buckley** design, Dan Pearson, **bl GAP**: Carole Drake, design: Gillian and Dick Pugh, **br Jonathan Buckley**; **124-125 Yann Monel**, Ballue; **127tl GAP**: Richard Bloom, Cogshall Grange, design Tom-Stuart Smith **tr Jason Ingram**, **b Clive Nichols**; **128t GAP**: Howard Rice, **b GAP**: Jonathan Buckley, design Carol Klein, **131t Emily Daw**, Lower House Garden, **b MMGI**: Marianne Majerus, Le Domaine viticole Claude Bentz, Luxembourg; **132-133 GAP**: Jerry Harpur; **134-135 GAP**: Richard Bloom; **136 MMGI**: Marianne Majerus; **139 Sabina Ruber**; **141tl and tr GAP**, **b GAP**: Sharon Pearson; **142 MMGI**: Bennet Smith, Great Dixter, East Sussex; **145tl GAP**: Martin Hughes-Jones, **tr GAP**: Rob Whitworth, **bl GAP**: Jonathan Buckley, **br GAP**: Tim Gainey; **147tl GAP**: Nova Photo Graphik, **tr GAP**: Jason Ingram, **cl GAP**: Jonathan Buckley, **cr GAP**: Carole Drake, **bl Naomi Slade**, **br GAP**: Jason Ingram; **148-149 GAP**: Rachel Chappell, Welford Park; **150-151 Clive Nichols**, Hampton Court Gardens, Herefordshire; **152 Jason Ingram**; **155tl GAP**: Jonathan Buckley, Perch Hill, design Sarah Raven, **tr GAP**: Gary Smith, **b MMGI**: Marianne Majerus; **156tl Jason Ingram**, **tr GAP**: Heather Edwards, **bl GAP**: Friedrich Strauss, **bl Clive Nichols**; **159tl Jason Ingram**, **tr and bl MMGI**: Marianne Majerus, **br GAP**: Heather Edwards; **160 Jason Ingram**; **162-163 MMGI**: Marianne Majerus, design Julie Toll; **164l GAP**: Lynn Keddie, **r GAP**: Mark Bolton; **166-167 MMGI**: Marianne Majerus, design Julie Toll; **168-169 GAP**: Jo Whitworth, Knoll Gardens; **170 GAP**: Jonathan Buckley; **173tl GAP**: Clive Nichols, **tr**

MMGI: Marianne Majerus, **bl GAP**: FhF Greenmedia, **br GAP**: Jacqui Hurst; **174 GAP**: Maddie Thornhill; **177tl and b MMGI**: Marianne Majerus, **tr GAP**: Carole Drake; **179tl GAP**: Jonathan Buckley, Ashwood Nurseries, **tr GAP**: Annie Green-Armytage, **bl GAP**: Nicola Stocken, **br GAP**: Martin Hughes-Jones; **180tl Jason Ingram, b Jason Ingram**: RHS, **tr MMGI**: Marianne Majerus; **183 Mette Krull**; **184-185 GAP**: Friedrich Strauss; **186-187 Clive Nichols**, Drummond Castle, Scotland; **188 GAP**: Heather Edwards; **191 GAP**: Howard Rice, Cambridge Botanic Gardens; **192tl and tr MMGI**: Marianne Majerus, **cl GAP**: Marin Hughes-Jones, **cr GAP**: Carole Drake, **bl GAP**: Matteo Carassale, Bembuseto, **br GAP**: Paul Debois; **194-195 Ray Cox Photography**; **197tl MMGI**: Marianne Majerus, **tr GAP**: Richard Bloom, **cl GAP**: Jacqui Dracup, **cr GAP**: Howard Rice, **bl Clive Nichols, br GAP**: Martin Hughes-Jones; **199 GAP**: J S Sira; **200-201 GAP**: Nicola Stocken; **202-203 Jonathan Buckley**, Upper Mill Cottage, Kent; **204 Jonathan Buckley**, Ashwood Nursery, Staffordshire, design: John Massey; **207t GAP**: Joanna Kossak, Sir Harold Hillier Gardens, **b MMGI**: Marianne Majerus, Sylvie Scweig; **209 David C Phillips**; **210 GAP**: J S Sira; **212-213 MMGI**: Marianne Majerus, St Timothee, Berkshire; **214-215 MMGI**: Marianne Majerus, Sussex Prairie Garden, West Sussex; **217tl GAP**: Lynn Keddie, **tr GAP, cl and bl GAP**: Andrea Jones, **cr GAP**: Sharon Pearson; **br GAP**: Tim Gainey.

Cover images: Front: **Alamy Stock Photo:** Tamara Kulikova

All other images © Dorling Kindersley

Author's Acknowledgments

This book is dedicated to my father, Roger Slade, in our mutual appreciation of a cool damp season and its quietly beautiful plants.

I would like to thank all those who have helped in the creation of *The Winter Garden*. It is a project that I carried in my mind for some years, and it is a great pleasure to see it so beautifully realized.

I must thank Rae Spencer-Jones and Helen Griffin at the RHS, and also Chris Young for his support and consistency. It has been a delight to work with the team at DK and my thanks go to Amy Slack, Ruth O'Rourke, Barbara Zuniga, and Maxine Pedliham for bringing their skills to bear at every stage of the mission. I am grateful to Diana Vowles for her editing skills, to Nikki Ellis for her sharp design eye, and particularly to Emily Hedges for her uncanny ability to read my mind and pick the perfect picture.

I am also indebted to Maggie Last and Joe Sharman for their thorough read-through and valuable feedback, to the owner of every inspiring winter garden that I have ever visited, and also to the RHS Shows Team, whose commission of a garden installation based on the Wisley Winter Walk crystallized the concept still further.

Finally, I would like to thank my family and friends for their encouragement and perspective, and particularly Chris and our children for their tolerance and humor, in the face of my unremitting passion for plants.

Publisher's Acknowledgments

DK would like to thank Adam Brackenbury for repro work, Mylène Mozas for design concept work, Francesco Piscitelli for proofreading, and Vanessa Bird for indexing.

Naomi Slade is a garden expert, journalist, author, and consultant, with over 25 years' experience. With a background in biology and a passion for conservation, she enjoys seeing plants in their natural environments and exploring the flora of different climates, while in her professional capacity, she has visited gardens, nurseries, and flower shows across the UK and Europe.

An award-winning show garden designer, Naomi was named Practical Journalist of the Year by the Garden Media Guild in 2022. She has also presented on BBC Two's *Gardeners' World*. Her previous books include *The Plant Lover's Guide to Snowdrops* and *An Orchard Odyssey*, together with *Dahlias*, *Hydrangeas*, *Lilies*, *Lilacs*, and *Ranunculus*.

For more information see www.naomislade.com, or follow Naomi on Twitter @NaomiSlade and on Instagram @NaomiSladeGardening

Project Editor Amy Slack
US Editor Sharon Lucas
US Consultant John Tullock
Senior Designer Barbara Zuniga
Production Editor Tony Phipps
Production Controller Rebecca Parton
Jacket Coordinator Jasmin Lennie, Abi Gain
Editorial Manager Ruth O'Rourke
Art Director Maxine Pedliham
Publisher Katie Cowan

Editorial Diana Vowles
Design Nikki Ellis
Picture Research Emily Hedges
Consultant Gardening Publisher Chris Young

ROYAL HORTICULTURAL SOCIETY
Consultant Simon Maughan
Publisher Rae Spencer-Jones, Helen Griffin

First American Edition, 2023
Published in the United States by DK Publishing
1745 Broadway, 20th Floor, New York, NY 10019

A catalog record for this book
is available from the Library of Congress.
ISBN: 978-0-7440-8441-2

Printed and bound in China

For the curious
www.dk.com

MIX
Paper | Supporting
responsible forestry
FSC™ C018179

This book was made with Forest Stewardship Council™ certified paper - one small step in DK's commitment to a sustainable future. **For more information go to www.dk.com/our-green-pledge**